DATE DUE

Pat Spallone is a freelance researcher and writer associated with the Centre for Women's Studies at the University of York. She previously worked as a biochemist in medical research at the University of Pennsylvania School of Medicine in Philadelphia, Pennsylvania in the USA. She is the author of a book on new reproductive technology entitled *Beyond Conception: The New Politics of Reproduction* (Macmillan, 1989), and was co-editor of *Made to Order: The Myth of Reproductive and Genetic Progress* (Pergamon, 1987).

PAT SPALLONE

GENERATION GAMES
Genetic Engineering and the Future for Our Lives

T TEMPLE UNIVERSITY PRESS
Philadelphia

To Linda Bullard

Temple University Press, Philadelphia 19122

First published in Great Britain by The Women's Press Limited 1992

Printed in Great Britain

ISBN 0-87722-966-X (cloth)
ISBN 0-87722-967-8 (paper)

CIP data available from the Library of Congress

Contents

Permissions

I would like to thank the following for their kind permission to reprint material in this book:

British Psychological Society
Cambridge University Press
Council for Responsible Genetics © geneWATCH, the bulletin of the Council For Responsible Genetics, 19 Garden Street, Cambridge, MA, USA
Dag Hammarskjold Foundation (the quotes are from *Development Dialogue*, the journal of the Dag Hammarskjold Foundation.)
The Ecologist (Reprinted with permission from *The Ecologist*, Station Road, Sturminster Newton, Dorset DT10 1BB.)
Greenwood Publishing Group (Reprinted by permission of Greenwood Publishing Group, Inc., Westport, CT from *Immunization: The Reality Behind the Myth* edited by Walene James. Copyright © Walene James 1988)
Guardian
Harvester Wheatsheaf,
Medical Research Council
New International (New Internationalist Publications)
Pluto Press (Reprinted with permission of the publisher Pluto Press Ltd)
Penguin
Pergamon
Victor Gollancz Ltd

And I would also like to acknowledge the following sources:

BBC for the *Horizon* script of the TV programme *The Book of Man*
Beech Tree Books, for *Gene Wars: Military Control Over the New Genetic Technologies*, Charles Piller and Keith R Yamamoto
Bulletin of the Atomic Scientists, © 1983 by the Education Foundation for Nuclear Science 6042, South Kimbark, Chicago, Illinois 60657 USA, for 'Recombinant DNA and Biological Warfare', Susan Wright and Robert L Sinsheimer
Doubleday, New York, for *The Ethics of Genetic Control: Ending Reproductive Roulette*, Joseph Fletcher
Farrar, Strous & Giroux Inc, for North American rights for Susan Sontag, *AIDS and Its Metaphors*
George, Susan, for US rights to *A Fate Worse Than Debt*, Susan George
HarperCollins for *The Female Eunuch*, Germaine Greer
Idoc Internazionale, for 'Biotechnology – Where to Now?' number 2, special issue; March/April 1988
Macmillan for *The International Directory of Biotechnology 1985*, J. Coombs
Marshall, J Richard, for 'The Genetics of Schizophrenia Revisited', *Bulletin of the British Psychological Society*
MRF (Malte Rauch Filmproduktion), for *Monkey Business: The Myth of the African Origin of AIDS*
The Nation, The Nation Company, Incorporated © 1983 for 'DNA – Key to Biological Warfare?', Charles Pillar
New York Times Magazine, for 'The Total Gene Screen', Morton Hunt
Pantheon, for American rights to *The Making of the English Working Class*, EP Thompson
St Martin's Press, for North American rights to *Genetic Engineering: Catastrophe or Utopia?* Peter Wheale and Ruth McNally
Wallace, Marjorie, for 'Unlocking A Family Secret', in the *Times*
World Resources 1988–89, World Resources Institute and the International Institute foe Environment and Development in Collaboration with the United Nations' Environment Programme, for UN statistics.

Acknowledgments

Deborah Lynn Steinberg read an entire rough draft of this book, at least twice as long as the finished version, and how much effort it took, only she knows. I cannot possibly thank her enough for her assistance, encouragement and valuable thoughts. Throughout the researching for and writing of this book, many people aided me in many different ways. As usual friends and those I work with around these and related issues have helped in more ways than they know, as have people I met at meetings, on courses, at bus stops and other such improbable places. Others supplied me with research material, insights and information when I found myself in conversation with them or sought their advice. Among these, I wish to thank in name: Paula Bradish, Linda Bullard, Rinki Bhattacharya, Sarah Franklin, the *Spare Rib* Collective, Alastair Hay (a few years ago now), Margret Krannich, Ruth McNally and Peter Wheale, John Newall, Mick and Jane Pythian, Helga Satzinger, Chayanika Shah, David Seedhouse, Mark Cheverton and Sayfu Ketema, John Lawton and Mark Williamson, and Women's Health Information Service (York). I also wish to thank the women and men I met while attending the FINRRAGE-UBINIG women's conference on health and reproductive engineering at BARD, Comilla, Bangladesh, 8–11 May 1989, who made the urgency and complexity of these matters all the more clear. May I take the opportunity to thank UBINIG (Policy Research for Development Alternative) for their warm and gracious hospitality. Also, my appreciation goes to the people I met while visiting laboratories in Hyderabad, Delhi and Bombay, some of whose names are cited in this book. Special thanks to Pushpa Bhargava of the Centre for Cellular and Molecular Biology for his special hospitality and for the time he spent with me talking about his vision of science and society. The British Council sponsored my flight to the Indian subcontinent to make these visits, for which I am grateful. I must again thank Chayanika Shah, for her friendship and care when I was in Bombay. At home, thanks to David White for all these things.

For the first time in my life I understand why so many book acknowledgments carry the disclaimer, 'Thanks to everyone who assisted me, I couldn't have done it without them, they saved me from many a *faux pas*, but all the mistakes are my own.' All the mistakes are my own, and I am totally responsible for how I represent organisations, individuals, social movements, other analysts' viewpoints and scientific terms and principles. I hope those whose names or associations are used in this book, whether they agree with my analysis or not, find that they have been represented fairly.

My thanks to the Centre for Women's Studies at the Institute for Research in the Social Sciences, University of York, for renewing my association as an honorary visiting scholar for the time I wrote this book; and to the British Sociological Association for their support.

Jo Campling believed in this book and got it off the ground. We were fortunate to find the support and enthusiasm of Ros de Lanerolle at The Women's Press, who opened a door for it to actually happen. Finally, this book would have had more errors and wobbly arguments had it not been for the care and attention of Hannah Kanter, Katherine Bright-Holmes, Judith Murray and Kathy Gale at The Women's Press, and for Lene Koch's and Alex Balsdon's exceptionally fine comments on the final draft.

Introduction

The idea for this book began in a peaceful setting. My friend Lene Koch and I were on a holiday together in the south of Sweden on the edge of a wood. One sunny afternoon, I was in the narrow sitting room of the house, reading books about genetic engineering and taking notes to prepare for a further education class I was going to tutor. Lene came in and we discussed how we had begun to think about the issues involved in genetic engineering with other women who were also concerned at the changes taking place in reproduction, agriculture, medicine and food production which are touching not only women's lives intimately, but those of every person and living thing. The issues are manifold and complicated: there is no one single judgment to pass on them. I was looking at the science as one trained in chemistry and biochemistry, and also as a feminist, as someone deeply affected by the social movements of my time. Lene said she would like to read a book about genetic engineering and biotechnology which took that focal point, and she suggested I write it.

Partly, my decision to embark on this project was made out of sheer shock at the mind-boggling scope of genetic engineering, and the wealth of concerned opinion and critical analyses coming from many different sources (but which does not, however, receive sufficient popular media coverage). I wanted to bear witness to both the developments and the concerned and critical opinion. Early doubts as to whether it was useful and responsible to explore genetic engineering as a subject of political analysis were soon gone. Genetic engineering is a politically important subject. It is considered one of the great transforming technologies of this century, destined to be a key technology of the next one, its goals ostensibly for the improvement of the world and its inhabitants. The scientific experiments of genetic engineering – in the sense of its widest possible meaning – are social experiments, in that they are affecting every level of our lives, from the food we eat and how it is grown, to health care and how it is administered, to the great expectations we have for our babies, to our perceptions of ourselves, other living things and nature itself.

The task of writing this book proved much more difficult than I had imagined. I felt overwhelmed with information: the technology was developing at great speed, and protests against genetically engineered products, against releasing genetically engineered organisms into the environment and against the patenting of them have been affecting the output and content of information from scientists, industries and governments about it. There seemed a million different critical analyses of a thousand different developments. How could I deal with all the material, on my terms – as the author responsible for the final shape and for the words of this book? One day, when I was feeling as if I were lost at sea, my friend Marilyn Crawshaw said, 'All we can do is raise questions.' I realised that that is what I had been doing, among other things, and this thought became a way to steady my course. Another thing that kept me going was my wish to write about a journey I had made without realising it while carrying out the research for this book, an account which became chapter seven, Genes-the-Cause, which covers some of the innumerable races to 'find' genetic causes for almost every kind of illness. This was the most important chapter for me to write, and it addresses concepts of 'genetic burden', eugenics and biological determinism.

The subject of this book is genetic engineering and in particular reproductive engineering. It is not an explication of methods by which frogs are cloned, gene 'fingerprints' are made, or mice are engineered to the size of rats. I do not wish to focus on the technology in isolation from the world, the whole context in which genetic engineering is used. I wanted to examine its origins in the history of science and in a broader social history, and its destination. I wanted to get beyond the usual presentation of the legal and philosophical ramifications of the science, beyond the view of genetic – or biological – engineering as either intrinsically anti-nature or intrinsically progressive. Legal and philosophical assessments of genetic engineering fluctuate constantly between these two poles. The one view is that the manipulation of life forms is artificial: either something to beware because it is going against nature, or something to be proud of because human beings are outwitting nature. The other view holds that the manipulation of life forms is not new and that there is nothing to fear and everything to gain because human beings have been altering nature and controlling the reproduction and heredity of plants, animals and themselves since the dawn of history. Both of these ways of seeing genetic engineering misread it, limit the discussion considerably and tend to obscure the economic and social factors which influence the development of this technology now.

In order to address all these aspects of genetic engineering, I had to consider it in relation to the wider subject of biotechnology. Biotechnology and genetic engineering are often used synonymously today, and there is a good reason for that. Genetic engineering techniques, developed in the late 1960s and early 1970s, form the linchpin of biotechnology, and have opened the door to the heralded 'bio-revolution'. Today, therefore, biotechnology is wholly concerned with controlling the characteristics of organisms by controlling and manipulating their genes, either directly or indirectly. However, I prefer to make a distinction between biotechnology and genetic engineering for the sake of clarity. I explain the distinction briefly here and in the glossary, and in more detail in chapter two.

Biotechnology is the exploitation of living things, and of substances from living things, to create products and processes for many different purposes. *Genetic engineering* is also concerned with all of the above, but not all biotechnological processes entail the use of genetic engineering techniques. For example, microbiological mining may use either 'naturally' existing – or genetically manipulated – strains of bacteria to enhance or induce acid leaching of metals from ores.

To complicate matters, the term *genetic engineering* has both specific and general uses. Any deliberate change of an organism in order to change its genetic characteristics might be called genetic engineering. Yet, the term genetic engineering did not emerge until recently, when researchers created methods to manipulate genes directly.

I always consider genetic engineering in the wider context of biotechnology, for both may be understood as the exploitation of living things and of substances from living things. I often use the phrase *biological engineering* to give an impression of the whole picture: genetic engineering in the broadest sense, meaning scientific manipulation of genes and genetic mechanisms, either directly or indirectly, with the aim of controlling the heredity or characteristics of an organism. By this I mean to convey, not only the scientific techniques involved in manipulating cells, genes and other parts of living things, but also the assumptions, social, economic and political, behind the techniques.

I want this book to do many things. Most importantly, it is meant to sow seeds for thought about genetic engineering:

thoughts about its implications for a new industrial revolution, for women's welfare, Third World development, disability rights and eugenics, food security, environmental welfare, animal welfare; thoughts about its ability to answer concerns about distribution and use of resources, about health care priorities; thoughts about the lack of context in the scientific model and the shortcomings of gene-think. In order to do so, I have presented data and arguments, not as the final word about a particular point, but to bear witness to both the developments and the controversies.

My thesis is that the relentless development of genetic, reproductive and molecular biologies today is implicated in a particular industrial, scientific and political agenda which is at loggerheads with progressive social development. Whatever the benefits that may come of it – and I think they will be small – they come at great cost. If one phrase could sum up the means of the bio-revolution it is 'the industrialisation of life', which I explore further in chapter five, Who Owns Life?, which discusses the patenting of living things.

In organising my critical thoughts around these developments as a *technology*, I am not presupposing that genetic engineers are the mad scientists among scientists. Any scientist working in the areas of biological and medical research may find themselves doing some sort of related work. Genetic engineering is not confined to glaringly obvious new developments, such as creating giant mutant mice implanted with human genes. I am starting from the understanding that a technology is more than just the machinery or the tools or the new methods of scientific innovation. A critique of a technology, such as nuclear technology or biotechnology, should include the whole context in which it exists; how it becomes integrated in our lives whether we want it to or not; the support systems and values it must incorporate; and its social reverberations – that is, how it affects people's lives and how the lives of some people are affected differently from others.

Such bold statements of intent and feeling may label those of us who voice them Luddites, the image connoting a disenchantment with technological change and an almost unthinking, knee-jerk reaction against machines, tools or new methods. Having been called a Luddite myself, I traced the history of the word, and found a meaningful origin in an account by the historian E P Thompson in *The Making of the English Working Class*.

The original Luddites were named after Ned Ludd, a defender of the rights of workers during a time of great social upheaval in England, namely that of nineteenth-century industrialisation. 'Luddism lingers in the mind as an uncouth, spontaneous affair of illiterate handworkers, blindly resisting machinery,' Thompson recalled, but their lives and experiences do not bear out this image.

The Luddite movement arose at a crisis point in industrial conflict. Its existence was brief (1811–13) and confined to specific areas and occupations in the north of England. It took the form of a carefully controlled direct action against the unrestrained forces of industrial capitalism: the break-ing of new machinery, Thompson explains:

> They were in direct conflict with machines which both they and their employers knew perfectly well would displace them . . . Despite all the homilies addressed to the Luddites (then and subsequently) as to the beneficial consequences of new machinery or of 'free' enterprise – arguments which, in any case, the Luddites were intel-ligent enough to weigh in their minds for themselves – the machine-breakers, and not the tract-writers, made the most realistic assessment of the short-term effects.

Between 1806 and 1817, as machinery for milling and shearing increased, skilled labourers found themselves out of work, or only partly employed, or displaced by lower-waged unskilled men and child workers.

Even if we make allowances for the cheapening of the product, it is impossible to designate as 'progressive' in any meaningful sense, processes which brought about the degradation for twenty or thirty years ahead, of workers employed in the industry.[1]

This understanding of Luddism as a specific response to a specific change is relevant to the resistance to genetic engineering: in both instances the question is, *what* is changing and *how*? Now when people voice fears or dislike of change, it often means technology-induced change: for instance, the use of computers in almost every aspect of public life. Such fear may be unfounded or founded in prejudice; however, it may be well grounded, even if not well defined. I hope to define and examine thoroughly the resistance genetic engineering raises in so many quarters.

'Luddites' are often seen as technological determinists: arguing that technology is 'bad'. In fact, if anything, it is the genetic engineering imperative which points to a technological determinism – the idea that innovation in biotechnology is absolutely essential at this time in history. For proponents of genetic engineering argue that molecular biology and its technology will solve all sorts of problems, from food scarcity to extinction of species to environmental pollution to infertility, and that it is the answer to our dreams of perfect babies, tastier food, cures for all ills, and life everlasting. Amen. It amazes me that the most relentless advocates of biological engineering imply that there is only one choice: embrace biotechnology or perish. It sounds awfully fundamentalist.

In the history of Western ideas and beliefs, determinism or fate was a concept placed in opposition to free will. I want to be clear that, in my critical analysis of genetic engineering, I am firmly of the belief that *choices are made*, although the making of a choice can be limited to accommodate a technology. We might well ask, 'How do these decisions get made?'

Further, although I am deeply critical of genetic engineering, I am not rejecting science as I understand it, precisely *because* it is so enmeshed in our lives. There are many things that happen within the category of science that are none the less social, human activities: problem solving, agriculture, medicine, the utilisation of the environment – all are not necessarily technology-based (organised, establishment) science in the contemporary sense. How scientists are trained and how they practise science are political issues. As are the multiplicity of opinion that exists and the parts that politicians, lawyers, judges, social scientists and doctors play within science. The language of science is common parlance these days; it's mass media.

Feminists like myself who suggest that patriarchy has something to do with the present state of scientific affairs, which are after all social affairs, are often accused of subscribing to a conspiracy theory. Strange, as genetic engineering and the 'revolution' in biotechnology are hardly secrets, but on the contrary are presumably legitimate activities enabled by legitimate leaders in science, industry, medicine and government. I am interested in examining genetic engineering at face value, and exposing the harm it is causing and will cause in the future. I will leave the matter of conspiracy theory to the lucid response of the writer Susan George, who said in her book on the Third World debt crisis, *A Fate Worse Than Debt*:

> I DON'T BELIEVE IN THE CONSPIRACY THEORY OF HISTORY. I take special pains to state this, because I've been accused of just such beliefs the moment I pointed out that a great many forces were converging in a single direction. They don't have to conspire if they have the same world view, aspire to similar goals and take concerted steps to attain them.[2]

Among my aims in this book are, to trace some of the converging economic, scientific and social – that is, political

– forces regarding genetic engineering; to examine the economic model from which it emerges; and to examine the *specific* goals behind the general promises of health and prosperity through biological engineering.

Why do this as a feminist? Feminists have been challenging the assumptions behind mainstream critical analyses of genetic engineering, thus opening up avenues of discussion which have been glossed over, for some time now. By the early 1970s, the prospects of the 'new genetics' raised a host of critical voices within science and outside it. In 1973, in a book entitled *Genetic Fix*, Amitai Etzioni concerned himself especially with direct human applications of what he called 'the foundations of tomorrow's technology', some of which were mass screening for 'carriers' of genetic disease, birth control pills, test-tube babies, controlling the sex of babies, using children and adults for experiments in medical science. These issues remain with us today, but have never quite been integrated in the discussion of genetic engineering as a whole, other than by feminists.

Perhaps the most neglected issue is that of the relationship between biotechnology and a perceived population problem (in terms of quantity and quality). Not a month goes by that someone somewhere is not blaming the planet's woes on over-population now or the prospect of it in the near future, which amounts to blaming *certain* people for there being too many human beings, or too few as in the case of racist neo-Malthusians worried about the decrease in birth rate among white populations. I have heard environmentally minded critics of genetic engineering suggest that we need not genetic engineering, but population control to solve the food and energy crises. Such a statement fails to address the fact that the world's food and energy resources are not distributed equally among the people of the world; and that not all people exploit nature the way industrial processes do. It is such reactionary shortcomings that this study informed by a feminist sensibility seeks to challenge and overcome.

In writing this book, I relied on different sources of information. Some was oral – from discussions, listening to interviews, conference presentations and speeches. I often relied on media information, which is a way of keeping up with the day-to-day developments in the fast-changing field of biotechnology. I travelled to India to see what was happening in research laboratories there. I read science and medical literature, institutional reports, and the books, articles and reports of other writers and analysts. I talked to women who share these concerns. I learned from the many books on biotechnology which cover similar ground to this one, and often very well. In particular are two superb and incomparable issues of *development dialogue*, which is published by the Dag Hammarskjöld Foundation. The first issue, appearing in 1983, was written by Pat Roy Mooney; the second, by Cary Fowler, Eva Lachkovics, Mooney and Hope Shand, all from the Rural Advancement Fund International, a small non-profit organisation. From these researchers and activists, and others like them who have written in this area, I have learned. Much of their background information and analyses are used in this book.

I have organised the book into two broad divisions: non-human genetics and human genetics, but have not kept to strict boundaries in my discussions of issues within these broad categories.

First, chapter one lays the groundwork by describing the rudiments of the new technology – what genes and genetic engineering are, and some genetic principles; and then by locating genetic engineering in the economic context of the bio-revolution.

In chapters two through five, the technical developments of non-human genetics – the manipulation of bacteria, viruses, plants and animals – are explained in terms of the three developments which became the focal points of the debate in Europe on genetic engineering.

● The first is the animal welfare and public health

implications of BST, genetically engineered bovine growth hormone. BST provides a case study in chapter two, which is concerned with the nature of the biotechnology industry, and which includes another miniature case study, genetically engineered 'human insulin'.

● The risk of releasing genetically engineered organisms into the open environment is the subject of chapter three, which discusses the weakness of present safety measures and several cases of legal and illegal releases of mutant viruses and bacteria.

● The changes in patenting law to accommodate the patenting of any living thing or parts of living things are the subject of chapter five which bridges the division between human and non-human genetic engineering.

The entire book is concerned with global and Third World development, but chapter four discusses the issues at length by placing the Genetic Supply Industry, the take-over of all aspects of food production by a few transnational companies, in the historical context of the Green Revolution.

Most of the second half, chapters six through nine, is devoted to human genetics: concepts about genetic disease, genetic explanations for human characteristics and behaviour, genetics in human reproduction, and beyond to the implications of 'genetic profiles' and the project now taking place to 'map' all the human genes. Chapter six, Future Ills, became my response to the promise that genetic engineering will bring great medical cures and products such as a vaccine for AIDS, and it lays the ground for the next chapter on human genetics in medicine. I found these explorations the most illuminating of all the work I have done on genetic engineering, for they showed me a progression from the old germ theory of disease to a gene theory. Genes are increasingly becoming the perceived cause of all sorts of illnesses; whence genetic technology becomes the solution. Chapter seven, Genes-the-Cause, takes that idea forward, inspecting the genetics of illnesses as diverse as Huntington's chorea, schizophrenia and chemical poisoning.

Chapter eight turns to human reproductive technologies which are rarely perceived in the same light as other biotechnologies, or as other industries. Could this be because they most immediately affect women? This chapter was the most difficult to write, because I have already written a book on the subject of 'new' reproductive technologies. I wanted to say everything I had said then again and more, but settled for concentrating on the importance of genetics in reproduction technology today.

Chapter nine is divided into two parts. The first looks at the ultimate human gene analysis project, the 'human genome project', aimed at mapping the genes on the chromosomes and sequencing all the human genes (deciphering their chemical formulae) with the presumption that genes will provide a book of revelations about humanity. The second tells the story of DNA 'fingerprinting'. If the genome project has aroused the most controversy, DNA fingerprinting has aroused least and, indeed, has arguably been put to good use in some cases.

Lastly, chapter ten examines the militarisation of biotechnology: the revived interest in and potential for biological warfare that genetic engineering arouses. I use this subject as a springboard to a wider discussion of the risks and dangers of all biological engineering, not just that used for military purposes, and of the morality of it – what has been called *gene ethics*.

The epilogue summarises my thoughts on the politics of the issues which arise out of genetic engineering, and some parting thoughts on gene ethics.

The subject of gene therapy, that is genetic engineering of human cells, became an appendix to the book for two reasons. The details are too important to leave out, yet the chapters where the discussion would have fitted in were already involved enough. Also, since the prospects of gene therapy and genetically engineered human beings crop up in several different chapters in the second half of the book, as an appendix it may be more easily accessible to the reader.

I have tried to explain scientific concepts (DNA, gene manipulation, etc.) as simply and accessibly as I know how, as and when they arise. In a few instances, detailed explanations may be given several pages after a term is first used. The glossary at the end of the book summarises key definitions for easy reference. I should stress that I have not discussed in much detail the discipline of immunology and the methods of cell fusion which are just as important to the biotechnology industry as are gene manipulation techniques.

Much has been left out of this book: I can think of significant developments on the subject in countries such as Australia, Brazil, Ireland, among others. I emphasise agricultural and medical applications of the science to the exclusion of others because I chose to concentrate on areas whose many facets I know best, and which illustrate the issues at stake best. And I am sure there are huge gaps in my knowledge.

Engaging in this debate has meant sometimes adopting a conceptual framework whose validity I question. One concept I am most sensitive about is that which divides the world into the so-called Third World or developing countries or the South, in contrast to First World, developed countries in the North or West. The countries thus lumped together under one category are diverse, as are the people who live in them. The geographical designations of 'North' and 'West' used to identify the industrialised world are inaccurate, as my friend Christine Crowe points out, since Australia is in the South geographically but economically is a highly industrialised nation and at the forefront of biotechnological innovation. But some words are needed to express the division which economic forces have shaped. Instead of consistently sticking to any one set of terms, and thus any one set of limitations, I've used all of them.

Biotechnology is a global issue, and it cannot be assigned such attributes as positive, negative or neutral, as delegates to the Dag Hammarskjöld Seminar, 'The Socionomic Impact of New Biotechnologies on Basic Health and Agriculture in the Third World' attested in the Bogève Declaration of 1987. 'Like other technolgies', they wrote, 'it is inextricably linked to the society in which it is created and used, and will be as socially just or unjust as its milieu.'[3] The meeting at Bogève, France, was concerned with two major areas of biotechnology: 1) the structural transformations in science, economics and politics which accompany the biotechnology revolution, and 2) the safety of biotechnology to human health, the environment and the food chain. Their concerns are the concerns of this book, as they are the concerns of the many people's organisations whose work, admittedly, 'has been reactive: serving as an early warning system of destructive policies and practices . . . But people's organizations have a prophetic function as well, as the Bogève Declaration attests, reasserting the primacy of human values in determining public policies'.[4]

1 The Gene Revolution

> We are moving from the Gold Standard to the Gene Standard. If ever you needed to understand at least the rudiments of a new technology, that time has come.
> Cary Fowler, Eva Lachkovics, Pat Mooney and Hope Shand, 'The Laws of Life'[1]

As the 1980s unfolded, genetic engineering came of age. Perhaps its effects can be summed up in a sentence from the report of the Federal Republic of Germany's Committee of Enquiry on Prospects and Risks of Genetic Engineering: 'Today genetic engineering itself is well on the way to becoming a technology for specific control of nature and of man.'[2] What kind of control, how, why, by whom and to what benefit it would be exercised all became important questions about the new technology, or, as some insisted, the not-so-new technology. The scope of genetic engineering, it became clear, was colossal.

Genetic engineering appeared everywhere, in industry, research sciences, agriculture and food processing, animal breeding and animal husbandry, medical clinics and medical research, police forensics and analytical chemistry, computer technology, space technology and military research. Genetically engineered products were on the market; genetically engineered bacteria, viruses and plants were being tested in laboratories and in open fields. Genetic engineering was being used to create more varieties of genetic screening, a new kind of gene therapy, genetically engineered animals, plants, agricultural inputs – seeds, bacteria, viruses – and the prospect of designer bioweapons. The message: it is the dawn of a revolution, a revolution in biotechnology which will bring genetic solutions to social problems (although military interest in biological engineering is never mentioned in the same breath as this message). The promises run like this:

- genetic engineering will make possible a new era in farming and animal husbandry, feeding the people of the world, and will be especially welcomed in famine-ridden countries
- genetic engineering will enable the production of designer foods, like leaner pork, for the health-conscious West
- genetic engineering is an environmentally sensitive technology: it will create biodegradable pesticides and ecologically sound fertilisers; the genes of bacteria and viruses will be manipulated, turning them into armies to protect food crops and trees from caterpillars and other pests; 'green' (environmentally friendly) enzymes manufactured by genetically manipulated fungi and bacteria will be set to work in paper and fabric processing, replacing polluting and resource-wasteful processes
- genetic engineering will create bacteria to consume oil spills in the oceans, and to make new plastics or living computer chips that will drive information faster than ever before
- genetic engineering will lead to vaccines and gene therapies that will wipe out diseases of all kinds
- genetic engineering in association with reproductive technologies will eliminate genetic diseases, thus allowing us to choose what kind of babies to have

- biotechnology will allow the world to *adapt* to the present-day social and environmental crises, 'to leap even further on to the "Man as God" bandwagon; to decide that we have, quite naturally, evolved the ability to genetically engineer alternative foodstuffs and artificial trees'.[3]

The Bio-revolution

The reason for setting biological engineering to these tasks is not simply because a scientific capability to do so exists: it has to do with the present economic climate, with scientific, industrial and economic stimuli, and with political choices made to enable this kind of technology. It has also to do with assumptions about the nature of social problems and their possible solutions.

Biotechnology was seen as the answer to the energy crises of the 1970s, to the depression of the petrochemical industry, and to the global recession. The scientific promise of biotechnology is directly related to its potential to restimulate economies, nationally and internationally, and as such it is seen as the saviour of the stock market. Hence, the interest of many governments, corporations, investors, scientists, advertisers and private medicine (the sale of sex-selection tests is a prime example).

It is useful to understand the background to this view of biotechnology. Fossil fuel resources are finite and dwindling. In addition, control over coal and oil resources is a political headache for governments and industrialists. The energy crises in the 1970s brought an end to cheap fuel for consumption in the industrialised world; oil prices rose and many of the Arab countries which were its major producers tightened control over the source. In addition, the products of the chemical industry are not as attractive from an industrial viewpoint as they once were: patents are running out of many of them; and the dangers of chemical pesticides are making them more difficult to market.

From this perspective, immense amounts of capital, research finance and resources have been and continue to be directed to biotechnology. In 1988 the US biotechnology enterprise was 'roughly equally underwritten by the federal government, which spends $2700 million annually on basic and "generic applied" research, and private industry, which invests nearly $2000 million.'[4] By 1989 the world's fourth biggest chemical group, ICI, was on its feet after a major slump. Like other transnationals taking advantage of the new biotechnologies, the UK company expanded its product areas and markets 'into just about every other production or consumer activity, with large divisions in pharmaceuticals and agrochemicals'[5] – the two areas which were feeling the earliest effects of the biotechnology boom. (However, in 1991 ICI profits were back at low mid-eighties levels owing to the economic recession.)

In a *Guardian* article (6 February 1991), Ben Laurance wrote that the company felt it was '. . . not here simply to clean up the environment,' but '. . . to conduct operations for the benefit of shareholders', which raises a question about the promise of biotechnology as environmentally and socially more friendly than the chemical industry. 'Biotechnology will have strengthened its position in the oleo-chemical, pharmaceutical and agribusiness areas, but as a partner not a threat to chemical technology,'[6] confidently stated Sir Geoffrey Allen, formerly at ICI, director of research and engineering at the Unilever Corporation, and a fellow of the Royal Society.

Similarly, governmental enthusiasm over renewed economic growth through biotechnology – along the same economic model in which the chemical industry's pesticide business flourished – suggests the limitations of calling it a revolution at all.

The promise of biotechnology as a *cheap*, efficient, clean answer to everything from the food crisis to medicine, when huge amounts of money are being spent on research and development in the hopes of profitable markets and Nobel Prizes (which have been awarded to many of the inventors of modern genetics), has to be questioned. Will the arithmetic work out in the long run? The investors hope so, and the prize winners need not care.

Transforming Technology

In some ways, however, biotechnology is revolutionary. Genetic engineers alter the heredity of living things in ways never before possible, beyond the boundaries of traditional breeding and cross-breeding methods, and beyond the boundaries of previous scientific capabilities. In the past, it was impossible to cross a rat and a mouse, or a tree and a petunia, or a person and a bacterium. Now it is possible, and it happens every day in laboratories and factories all over the world.

To put it another way, genes are being swapped between species. Human genes are put into animals and bacteria; rat genes are put into mice that grow to be giants; sheep and goat genes are mixed up with one another and result in a conglomerate animal. A species is defined as a group of individual organisms which can interbreed with each other. In the hands of biological engineers, that boundary line between species is transgressed, in the process of manipulation of the hereditary material in bacteria, viruses, humans, animals, insects and plants which never before has been possible.

The 'logical' outcome is the manufacture of totally new species and the development by genetic engineers of an 'alternative evolution'. One palaeontologist ventured into that future, probably saying more than he meant to in the title of his book *Man After Man: An Anthropology of the Future*. His illustrator drew an imaginary genetically engineered forest dweller, *Homo silvis fabricatus*, as a man, of course, hairy like an ape, with a human face, clinging with his ultra-elongated toes and fingers wrapped around the trunk of a tall tree, his head of long hair slicked back off his forehead, looking as if he faithfully took comb and gel to his locks every morning , à la Michael Douglas.[7]

However, despite the mind-boggling implications of the technological capability of crossing species boundaries, the revolutionary nature of biotechnology is not simply reflected in the potency of such 'technical' capability, nor is it simply contained in its promise of 'genes-the-answer', the all-encompassing solution to every problem, which may or may not come to fruition. It is more a matter of where biotechnology has come from, and where we are going with it, and of its effects on living things, including ourselves, and the world we live in.

Never before have living things been utilised as a pool of raw materials to this extent, like so many gold mines or oil deposits, to create products and methods of production for industry, medicine, agriculture and science. Biological engineering enables scientists, and ultimately those who fund them, governments and multinational companies, in their enthusiasm for biotechnology, to make claims on living things as exploitable natural resources, rather as nuclear physicists tapped into the heart of inanimate matter. Consider the scope of biotechnology – and the techniques of genetic engineering which make it possible as one of the great transforming technologies of the twentieth century, akin to nuclear technology or information technology: large institutions are developing it, marketing it, organising everything from food production to reproduction around it. It is, we are told, the Age of Biology, which heralds (supposedly) a future of health and prosperity.

Genetic engineering is not just a continuation of what human beings have done for aeons in plant and animal breeding, as its apologists often suggest. Only since the mid 1970s has biotechnology become regarded as a distinct scientific discipline. Only with the technological capabilities afforded by gene and cell manipulation were previously impenetrable species boundaries penetrated. Only in the 1980s was its economic potential appreciated.

'The potential of biotechnology is such that it will have dramatic effects in most areas of our lives within the next decade,' *The International Biotechnology Directory 1985* predicted, 'it will influence changes in raw material use, as well as patterns of labour, health, energy and food production. To realize this potential will require cooperation between nations with the international organizations playing an important role in such areas as the safety aspects of genetic engineering, patent protection, establishment of culture collections, seed banks and germplasm collections, regulation of food supplies, aspects of community health and the promotion of information exchange and joint research projects.'[8]

The transformations of medical care, food and industrial production and reproduction have only just begun, and they are impinging on many political and social struggles: social equality, reproductive autonomy, global development, disability rights, environmental and ecological concerns, animal welfare, agriculture and medical ethics, population control. The directors of this 'big' technology are also big: it is those with an industrial eye to the economics of dwindling resources – members of governments, corporations and international development and financial institutions – scientists, engineers, economists, demographers and policy makers who are taking us into the age of biology.

'The More Things Change. . .'

The bio-revolution is therefore not going to change much because there is no sense in which biotechnology proposes to address the root causes of the social injustices and inequalities which result in the crises the new technology is supposed to solve. There is no proposal that either resources, food, or energy will be reallocated within states or redistributed globally, strategies which would have far more impact on world energy and food crises.

The myth of food scarcity, for example, is dismissed in the report *World Resources 1988–89* by the World Resources Institute and the International Institute for Environment and Development in collaboration with the United Nations Environment Programme. It estimates:

> Hunger persists because food is distributed unevenly. If the amount of food produced in 1985 were distributed evenly throughout the world, it could provide an adequate diet for 6 billion people – 1 billion more than the Earth's population. . . Access to food is correlated with income.

The report goes on to say that modern farming methods, with their overuse of pesticides and fertilisers, are well known to have caused environmental destruction and crop loss in many areas of the world; while the practice of monoculture, planting huge areas with a single crop of specific genetic make-up, have caused other problems. 'Now some Asian farmers are reexamining traditional crop diversification as a sounder approach economically and environmentally,' eschewing 'the top-down laboratory approach' to research.[9]

The more things change, the more they stay the same. It is likely that socially and economically, the same rules and forces as applied in the first industrial revolution will apply in what has been called the second. The biotechnology revolution is promising something for nothing. It promises health and prosperity without a reallocation of priorities, or a change in the ways of thinking and the behaviour of industrialisation. In fact, the flagship of biotechnology, genetic engineering, will lead the continuing expansion of

industrial activity. Biotechnology is a science with an industrial mind-set. What particular dangers and problems does this pose?

There is evidence of an awareness of the dangers and problems, and of the costs, and to whom. When feminists in Germany held their first national congress of Women Against Gene and Reproductive Technologies in 1985, 2000 women – an unexpectedly large number – attended. By the summer of 1988 in Europe, public interest groups were holding meetings and calling for bans or moratoriums on three of the most pressing developments of the 1980s, namely, the release of genetically engineered organisms into the open environment; the use of the genetically engineered hormone BST in dairy cows; and a new phenomenon that has arisen with the development of genetic engineering, the process of patenting living things as if they were refrigerators or toasters.

Some of the concerns, which I will discuss in the following chapters, may be broadly outlined as follows:

• Genetically altered micro-organisms pose risks to human health and to the environment. Scientists and regulators promise to keep that risk to a minimum, but our own history warns of the 'whoops' theory of risk assessment (something can go wrong – or perhaps already has – and those affected are left with the apology, 'Whoops, sorry, we made a mistake').

• The field of human genetics, like other fields of genetics, is an arena for the assertion of social principles, prejudices and values. Who shall decide which genes are good, who should breed and who should live? What will be the impact on people with those disabilities which are defined as genetic which could thus be eliminated? 'Having been assigned the highly legitimate designation of preventive medicine, medical progress is now being devoted to eugenics,' said Bernhard Claußen to the European Workshop on Law and Genetic Engineering.[10]

- As transnationals gain wider control of plant and animal resources, what effect will this have on development in countries already dependent on those same companies?
- Women are disproportionately affected. The biotechnologies which can alter and shape the fertility of the soil, of plants, of animals, are also being applied to women as part of medical approaches to reproduction and foetal therapy. Women, too, are taking on the burden of changes in agriculture and food production. 'The fact that women own less than one percent of the world's property has traditionally concealed the fact that they produce at least half of the world's food. In Africa women produce 60 to 80 percent of all food, yet in the new dispensations of grain, technology, cash and loans, men are the favored beneficiaries.'[11]
- Animal welfare is at risk. Animals are the subjects of further cruel exploitation for the sake of bioengineering, even to the extent of becoming patentable. Our relationship with all living things and how we conceive of ourselves is at stake.
- The ability to control the new technology is questionable. Many scientists and industrialists see existing controls as excessive and, in any case, find loopholes through which to escape them. Numerous illegal experiments have been reported. How many more have not? Regulations are often made by the authorities only to allay public fears and to woo public confidence.

What kind of answer does this technology offer? What price this revolution where freedom is the freedom of the market place, to buy and sell living things and parts of living things – genes, gene products, cells, biochemicals extracted from cells and bodily fluids, pieces of viruses, etc. What effects is this having on human rights when the question of 'Who owns my body parts?', at a time when organs for transplants are in demand, challenges medical ethics? What

effect does this have on the rest of living nature? Are all living things to be reduced to the property of patent holders, investors and biological engineers?

These are familiar questions and problems; genetic engineering is not a new science, but a new set of novel techniques. The significant difference is that a whole, different category – living things – is transformable and exploitable in ways and on a scale never before possible. The social reverberations are not new, and neither is the economic and scientific model from which the biotechnology industry springs, which tends to serve the same economic and scientific interests as did the development of the petrochemical and nuclear industries which have gone before. The set of 'novel techniques' has enabled the generation of this technology and its astounding implications for the future.

Genetic engineering – *as a technology* which requires complex support systems and the acceptance of its value – is intensely political and, like any other transforming technology, does not exist in a vacuum. And although the essence of genetic engineering is not in the act of splicing genes, an exploration of some of the rudiments of genetics and genetic engineering is a good place to begin an exploration of the operation of the technology in society. However, any attempt to present the concepts of genetics and techniques associated with it should, I think, start with exposing the difficulties of clarifying such concepts.

The Problem of Clarity

To understand a science it is worth knowing what the practitioners of it do and what they believe. For those who are interested, there are many books which explain the structure and chemistry of DNA, the special pieces of DNA known as genes and the scientific methods of genetic engineering. Peter Wheale and Ruth McNally's *Genetic*

Engineering: Catastrophe or Utopia? painstakingly describes the details and also explains some of the jargon of genetic engineering in its discussion of the less publicised problems that ride in the wake of the science. Another source aimed at the general reader, this time coming from the scientific establishment's point of view, is *A Revolution in Biotechnology* edited by Jean L Marx.[12] I will not rehearse such explanations here, but I will explain some of what the practitioners do and believe. I do not think that one must understand every single scientific fact and facet of biochemistry and molecular biology to understand the implications of biological engineering, just as you do not have to understand the equations relevant to generating nuclear energy in bombs or power plants to come to an informed opinion about the effects of nuclear technology.

I emphasise this because critical public opinion is often dismissed as uninformed about or misunderstanding the science. This attitude is like a house of cards. If a 'fact' is deemed wrong, the argument with genetic engineering comes tumbling down, and any worries about its consequences are dismissed with the demolition of the argument. (By contrast, when one accepts the scientific view, one is considered enlightened and informed whether or not one is conversant with the 'facts'.)

I remember a conversation with a geneticist in which I cited the search to find the gene 'for' cleft palate as disturbing in terms of its capacity for social engineering, the attitude it reflects about normality and of discrimination against people who do not fit into that preconceived normality. Is anything which is defined as a 'gene defect' to be wiped away with genetic screening? The geneticist huffed impatiently that cleft palate is not a genetic condition, ending the discussion on his own authority. Apparently, the right to be concerned about the organisation of medical services, women's reproductive lives and of the rights of people with disabilities is dismissable in such a case. (A few months after this depressing conversation, the

newspapers reported that a researcher had finally isolated the gene for a rare form of cleft palate; just within those few months, the 'new genetics' was becoming more established and assured.)

I wish to make a point now about the problem of clarity. By that I mean the tension between trying to describe clearly and simply a scientific fact or principle and doing justice to its uncertainties and nuances. I am sensitive to the limitations of the language of any professional discipline which encodes certain presumptions – here, those about biological science. Worse is the false message a false clarity can project. Perhaps one may comfortably say that all living things have genes, although in 1991 there appeared reports of the discovery of a possible new type of microscopic life form which was composed entirely of protein molecules, that is, without any genetic material whatsoever; or perhaps one may say that genes are biochemicals which carry information the body needs. But saying so leaves much left unsaid, known and unknown, about the action of genes. Further, other statements are more doubtful, such as the relationship between genes and human characteristics. Perhaps one can say that mice can now be engineered which carry a human gene in every cell of their bodies but left unsaid is the fact that the rate for successful genetic manipulation of the embryos of laboratory and farm animals injected with the foreign genes is between one and three per cent.[13]

There are, of course, infinite complications and unknowns in any attempt to clarify scientific principles in this or in any book. Scepticism for the scientific Word is healthy: yesterday's scientific 'fact' can be tomorrow's scientific error; today's scientific heresy can be tomorrow's scientific fact. This is not to say that we live in a vague, dark world where we will never know anything. It is to say that the superiority complex of science and science-based technology must be challenged. It is to say that science and technology may – have? – become a kind of religion for modern times. Which is not a very scientific thing to happen. What is certain is that scientific principles describe and explain in terms of a particular framework (the chemistry of a thing, for example) and particular premises. Science is never free of its time, history, its cultural perspectives and biases.

Let us begin to examine the material called genes, some genetic principles, and the 'gene hypothesis' on which the science rests.

What are Genes?

All living things have genes. Bacteria have genes, plants and animals have genes, and people have genes in almost all the cells of our bodies. Hair cells, muscle cells, blood cells, nerve cells, egg and sperm cells, all carry the chemical messengers known as genes.

Before the material existence of genes was identified, it was an idea. The word gene replaced the term 'unit of heredity' in the early twentieth century, shortly after the rediscovery of the work of Austrian cleric Gregor Mendel (1822–84), whose cross-breeding experiments with pea plants were destined to lay the foundation of the science of genetics. Mendel's discoveries overturned a common European belief that heredity – 'like begets like' – was transmitted through blood or other fluids; that the bloods of the biological parents mixed in the offspring. Mendel showed that the characteristics he was observing in pea plants, such as red or white flower colour, were inherited as discrete units, and do not mix and blend with other inherited traits in the offspring.

A well-known textbook, *Principles of Genetics*, explained Mendel's 'gene hypothesis' as such:

Contrasting characteristics, such as the red and white flowers in pea plants, are determined by something that is transmitted from the parents to offspring in the sex cells, or gametes [commonly, egg and sperm cells]. This 'something' is now called a *gene*.[14]

The biochemical links between the abstract concept of genes and the tangible, observable features of organisms were eventually formulated and properly defined within the scientific world by 1945. That biochemical link became known as the 'one gene, one enzyme' rule, which states that each gene is responsible for an enzyme (a protein that aids biochemical reactions in the body); the enzymes or proteins, in turn, are responsible for important functions to maintain body structure and metabolism.

Today, geneticists refer to the combination of genes of an organism as its *genotype*, while the *phenotype* is the observable characteristics of the organism as determined by the interaction of its genotype and its environment.

The idea of a *universal applicability* of the gene hypothesis is a significant one, I think. The question to be kept in mind throughout this book is, 'How far can you go with the gene hypothesis?' In humans, the association of heredity and characteristics in some rare instances seems straight-forward, such as that of blood groups. We inherit our blood group, and may be either group A, B, AB, or O. This is an example of a very specific instance of hereditary differences among people, but, as the late biologist and historian of science Sir Peter Medawar pointed out, few examples like this exist.

The identity of the 'something' known as a gene was clarified in 1953. Genes are the chemical DNA (deoxyri-bonucleic acid), the famous double helix, the emblem of genetic engineering. In plants and animals, DNA is packaged in long rod-like structures called chromosomes, located in the nuclei of cells. A bacterium, a single-celled organism, does not have such a nucleus; its genes are packaged in a main circular structure, and it also has extra genetic information in other circular structures called plasmids.

Viruses, although sometimes classified as micro-organisms, are not considered to be living things in the same sense as are bacteria, plants, animals and ourselves. Viruses are not composed of cells, and they do not have all the chemical elements necessary for reproduction within themselves. Viruses do, however, have genetic material, either DNA or a similar chemical called RNA (ribonucleic acid). And viruses do replicate, a sort of reproduction that can take place when viruses inhabit the cells of a 'host' organism, perhaps a bacterium or a human.

DNA is considered to be so fundamental because it 'contains the information that determines and controls the way cells, and thus organisms, function, grow and divide'.[15] However, this is a broad, general statement: what the biology and chemistry of DNA actually tell us is that genes are chemical 'messengers'. Genes *encode* for proteins. The chemical structure of a specific gene relates to the chemical structure of a specific protein. Genetic information is transcribed and translated in response to other chemical events and other chemical messengers (hormones, for example) in and around the cell. Genes do not act in isolation. External factors influence the working of genes in the cells and, indeed, external factors must often be altered to facilitate the objectives of a gene manipulation.

Furthermore, to say genes encode for proteins is also a simplification. Not all of the DNA is genes as such. In addition, theories of how genes work are constantly re-formed in the light of new evidence. But, for our purposes, the basic principle is this: a gene encodes for a protein. The proteins in turn give the organism its shape and carry out important functions. However, an important complication to the basic principle is this: there are a myriad of other factors existing in conjunction with genes, and the old rules of genetics constantly fall short. For instance, the standard rule 'one gene, one protein' is not immutable. Moreover, very little is known about genes and genetic mechanisms in complex organisms like humans, and environmental factors always contribute to how the body of any organism responds. Although the production of insulin or of growth hormone in our bodies depends on the instructions of a specific gene or genes (DNA), that is not the only thing they depend on. Although a virus or bacterium may be engineered to have a certain set of genes, they can and do mutate as they interact with their environments.

Gene Mutations and Recombinations

Two other important concepts of genetics which are carried over into genetic engineering are recombination and mutation.

Recombination is the rearrangement of genes in the sex cells (egg or sperm) in combinations that are different from what they were in the parent organism. During sexual reproduction, two sex cells from two different parents join, resulting in an offspring having a unique combination of genes. Traditional methods of cross-breeding for plant and animal hybrids rely on just such recombinations. Recombination can also mean any rearrangement of genes which occurs in nature (for example, the genes of a virus can integrate with human DNA in a human's cells), or in genetic engineering.

In molecular terms, a *gene mutation* is any change which results in an altered protein molecule. That change may be observed as altered biological activity; or there may be no noticeable change. Mutations may occur spontaneously or be induced by, say, environmental stresses such as cosmic radiation. More generally, a mutation is any change of a particular gene or chromosome structure. Mutation is an important concept in evolutionary theory, which states that genetic mutations which occur in nature, at a low rate, affect the appearance or behaviour of the cells and organisms involved.

Mutations can also be induced under laboratory conditions. Earlier in the century, the geneticist Hermann

Muller showed that x-ray radiation causes gene mutations in fruit flies. This discovery was considered to be a milestone in the study of genetics, and won Muller the Nobel Prize. However, any gene mutations induced by zapping an organism with x-rays are random – undirected and uncontrolled. By 1973 the ground had been laid for a whole new ball game.

Gene Manipulation

The term *genetic engineering* appeared in the mid 1960s as scientists increased their ability to manipulate the hereditary material. By the early 1970s, hybrid DNA (DNA formed from combining DNA from two different sources) had been created in a test-tube, and DNA foreign to bacteria was transferred into bacteria. Two of the research teams credited with the early discoveries and inventions of *gene manipulation* were located in California in the United States: under Stanley Cohen and Annie Chang at Stanford University, and under Herbert Boyer and Robert Helling at the University of California School of Medicine in San Francisco.

DNA exists as very long molecules. Genes are specialised regions located along a molecule. Researchers found a way to 'cut out' specific genes along the DNA by chemically splicing the long DNA molecules with a newly discovered class of enzymes known as restriction endonucleases or restriction enzymes. These were first isolated from the bacterium *Escherichia coli (E. coli)*. Restriction enzymes act like chemical scissors. They 'cut' a long strand of DNA in specific places. Another sort of enzyme can 'paste' DNA pieces back together again. With the aid of these chemical tools, scientists are able to isolate and recombine individual genes outside the organisms from whence they came.

Genetic engineers exploit other biological 'tools' as well. For example, the ability of viruses to enter cells, replicate inside them, and move along to other cells, has made them a useful tool for genetic engineering. A foreign gene may be added to the viral genetic material; then the virus can be used to introduce the foreign gene into the cells of bacteria, plants or animals.

To sum up, with the capability to 'cut and paste' DNA, scientists were able to synthesise novel DNA in the laboratory. Also, they could extract individual genes of interest from one source of DNA, for instance a human being, and paste them into the working DNA of another organism such as a bacterium. The new method of *gene splicing*, together with the other methods necessary to create a hybrid DNA, are known as *recombinant DNA technology*. The resulting product, the hybrid DNA, is known as *recombinant DNA*.

This kind of genetic engineering, recombinant DNA technology, arguably gave the impetus to the bio-revolution, attracting the attention of scientists, business people and the military to the field. Wheale and McNally summarise: 'The ability to decode and synthesize DNA molecules *in vitro* endowed the genetic engineer with the power to "read and write" in the language of genes.'[16]

Gene splicing or recombinant DNA technology is accurately described by Wheale and McNally as *micro*genetic engineering, since the genetic alterations are being made on the level of the DNA itself, and entail the use of methods of microbiology and molecular biology. (There are also other methods of genetic engineering and cell manipulation equally important to biotechnologists.)

In this chapter, I have attempted to lay the groundwork for the rest of the book, to state why the technology of genetic engineering is being introduced now, to get to grips with some of the rudiments of the technology, and to begin talking about its possible effects. Before moving on to the next chapters which focus on specific cases and issues, and in which I often draw heavily on the work of other researchers, analysts and activists, I should make explicit my relationship to these issues.

During the time I carried out the research for this book, it occurred to me that certain critical boundaries that I had once imagined useful no longer were. Is genetic engineering 'good science'? Is it 'bad science'? Or are the categories 'good' or 'bad' science not even useful here?

I think the latter, but I must admit that there are no neat categories in which to place the various assessments of genetic engineering. There are numerous critics who are saying, 'This is not good science'.

Of course, much depends on the definition of good science and whether there can be general agreement on this. Most important, I do not think we can possibly comprehend what 'good science' is or might be without taking account of the institutions, including the scientific establishment, which are engaged in the research of genetic ideas and genetic technology, and whose interests are served by them. This means taking account of power relations: who creates the technology, who directs its course, and how it impinges on our lives. A critique of genetic engineering that is, say, environmentally centred is inadequate if it does not take power relations into account, and any attempt at defining and enabling the welfare of living things which is insensitive to discrimination on the basis of sex, race, (dis)ability, economic or class status will continue to be oppressive and 'unscientific'. Philosopher Mary Midgley says that we must look at the content and methods of science itself, that 'We cannot leave ideology out of the picture'.[17] Consider, in particular, the fact that the cornerstone of modern science, the conceptual bifurcation of 'man and nature', is by many accounts the philosophy responsible for the scale of environmental degradation and human misery that exists in the world today.

Critiques which otherwise claim to be politically sensitive often assume and are informed by a misguided belief in universality; as if all human beings wield the tools of biotechnology, and as if all human beings are equally affected by biotechnology, thus masking the disproportionate effects on different groups of people.

On one momentous level alone, the very idea of organising society around gene-orientated ideas and practices should make us wary. The present women's movement in the industrialised world began with an awareness that sex roles and sex stereotypes – the supposed superiority of men and inferiority of women – were assumed to lie in our respective biological make-ups. The idea of biological superiority has been a deeply held belief in our world, a foundation of sexual subordination in the law and in our everyday lives. It continues to be difficult to challenge the social mores which embody the principle. It has been even more difficult to challenge the scientific mores which embody the principle. On the first page of the first chapter of Germaine Greer's landmark book *The Female Eunuch*, published in 1971, she ventured:

> It is true that the sex of a person is attested in every cell in his body . . . Perhaps when we have learnt how to read the DNA we will be able to see what the information which is common to all members of the female sex really is, but even then it will be a long and tedious argument from biological data to behaviour.[18]

How deeply suggestible is the science of sex difference.

Greer's prediction foreshadowed scientific interest which led to the discovery of the 'gene for maleness'. In 1990 Dr Peter Goodfellow's laboratory at the Imperial Cancer Research Fund discovered a human gene on the Y chromosome which appeared to determine sex: a gene which guides how the cells of the testes develop in the human embryo. It is called the SRY gene. A similar gene was found in mice. By experimenting on mice, Dr Robin Lovell-Badge at the National Institute of Medical Research, London, in collaboration with the cancer fund, succeeded in changing would-be female embryos to males. They injected female mouse embryos (newly fertilised mouse eggs) with the SRY gene. Three of the 11 genetically altered female embryos developed the physical characteristics of males and the expected sexual behaviour of male mice towards female mice, but they are sterile (the researchers explain that at least two more genes from the Y chromosome are necessary to complete sperm development).[19]

The identification of the 'key to maleness' or the 'genetic essence of masculinity', as the newspapers enthused, is considered one of the most important developments in genetics. It has considerable implications for animal breeding, medicine and knowledge of sexual development. The pig industry prefers male pigs, which are leaner and grow faster than females. Cancer researchers like Goodfellow are interested in learning about the relationships between the genes involved in development and 'cancer-causing' genes (oncogenes). As for fears of applying the sex-change method to humans, spokespersons for medical research insist that this will never happen. It would be morally repugnant and illegal. (It would also be an inefficient way to get rid of human females, if that were the sinister objective: once a female embryo is identified in the laboratory dish, if unwanted it could be chucked away.) However, it is not hard to imagine that in future researchers will argue that they must corroborate in human embryos what they have shown in animal embryos regarding, say, developmental diseases or theories of male sexual development. It would not be premature to ask what those future research concerns regarding 'male development' might be. Are we to believe that the essence of our sex can be found in our genes?

Now we are faced with, among other things, the 'human genome project', the project to 'map', to decipher all the human genes. Some of the molecular biologists carrying out the work take delight in expounding on how it will unlock the secrets of the human condition, how we are the same, and how we are different. How will such information be perceived and used? 'We cannot leave ideology out of the picture.'

2 Biotechnology Now

> Biotechnology is the ultimate chimaera, a hybrid of
> science and commerce, that forages for investment and
> opportunities worldwide.[1]

> I think BST is meant for doping cows.
> Wolfgang Goldhorn, veterinarian[2]

The ancient Greek myth of Prometheus tells us that human
beings have been altering their worlds in peculiarly human
ways since fire was discovered. Humans have utilised
plants, animals and the processes of invisible micro-
organisms like bacteria long before the advent of present-
day biotechnology and genetic engineering. Cheese-making,
baking and brewing are the examples most often given of
'biotechnology' in pre-history. In cheese-making, bacteria
convert milk into solid food. Brewing employs the fer-
menting properties of yeast, converting sugar into alcohol
and carbon dioxide gas, a discovery which is credited to the
French chemist Louis Pasteur (1822–95). The use of biological agents in war also precedes the discovery of disease-
causing germs. In the fourteenth century the cause of
plague, now known to be the bacterium *Pasteurella pestis*, was
put to military purpose when infected cadavers were hurled
over the walls of the besieged city of Kaffa on the Black Sea,

and eventually caused an epidemic.

So, why the controversy over present-day biotechnology? Isn't it just a more sophisticated version of what humans have always done? Or is biotechnology today appreciably different from the use of bacteria, plants and animals that has gone before? And what does each view imply?

This chapter begins with a brief trip back to the earlier biotechnology industry which emerged in the twentieth century, where we find some age-old processes in a brand-new world, pioneers and heroes, promising products with much enthusiasm. The chapter then moves into the present, to look at two of the earliest mass-produced products of genetic engineering: the cattle growth hormone BST which is being used on dairy cattle, and 'human insulin', an apparent medical breakthrough. Neither BST nor 'human insulin' lived up to its promise. The most troubling and resounding aspect of these two cases is the inability or unwillingness of their inventors, sellers and regulators to admit the now obvious shortcomings and dangers of the products and of the processes used to create them.

Biotechnology Now

Every book, article or industrial report on biotechnology produces a slightly different definition of what biotechnology is.

Biotechnology can be described as the use of living things and living processes to create products and processes for human use, industry, medicine and the military. The UK's Royal Commission on Environmental Pollution defined biotechnology as 'the application of scientific and engineering principles to the processing of materials by biological agents to produce goods and services'. Jean Marx in *A Revolution in Biotechnology* defined it as the use of living organisms, or substances obtained from living organisms, to make products of value for human beings. However defined, all agree that biotechnology will play an increasingly important economic role. Estimates of the market worth of biotechnologically derived products vary, but by the end of the 1980s one estimate put it at $100 billion worldwide by the year 2000, billions of that generated within the United States alone.[3]

The word *biotechnology* is fairly new. It does not have an entry in the 1933 edition of the Oxford English Dictionary, and it might have been first used in a speech in 1936 by Sir Julian Huxley, British scientist and eugenicist. Alluding to nineteenth-century industrialisation, Huxley predicted a new age of biology, saying:

> We falsely identified science with physics and chemistry, and technology with the technique of the early machine age. In reality, the machinery and the technology at present in use are for the most part crude and primitive compared with what might be achieved; biology is as important as the sciences of lifeless matter; and biotechnology will in the long run be more important than mechanical and chemical engineering.[4]

The word biotechnology has come into everyday language only in the past two decades when gene manipulation and cloning were added to the repertoire of scientific methods. Today biotechnology is often synonymous with genetic engineering, where genetic engineering is any deliberate change in the genes of an organism in order to change its characteristics. Some say that traditional animal and plant breeding could be considered as genetic engineering, since these are deliberate attempts at intervention and manipulation. However, as a term, genetic engineering made its appearance in the mid 1960s, and has come to mean the most recent technological interventions that enable scientists to introduce, delete, or otherwise change the DNA of an organism using the methods of molecular biology, cell biology and molecular genetics. They include not only gene splicing, as described in the previous chapter, but other methods such as cloning, splitting embryos in two and cell fusion (merging cells under laboratory conditions).

Therefore, although some commentators suggest that biotechnology has been around for millennia in the form of agriculture, and others cite cheese-making and brewing as earlier examples of biotechnology, the comparison at best is limited, at worst obscures the differences.

Biotechnology is as different to brewing beer in a kitchen as nuclear energy is to the burning of wood in a fireplace. There are similarities, but the technological interventions now possible, gene manipulation among them, have moved biotechnology into an entirely new realm, and one which is fraught with new and unknown hazards, compounding those already known, which are increased manifold by the staggering difference in degree and sophistication of biotechnology today. The new biotechnologies attract the support of big money and governments, and their commitment to the pursuit of biotechnologies means that other options fall by the wayside. A technology is not just a machine or a tool or a technique; it requires a support network founded on particular values to ensure its promotion.

Let us look at the first wave of the biotechnology industry, as it was before genetic engineering came on the scene.

Biotechnology Then

The biotechnology industry began with use of bacteria and yeast in the fermentation industry. Some historians claim it began in England in 1914, at the outbreak of the First World War. There was a shortage of the chemical acetone, which was necessary to make the explosive cordite. The Minister of Munitions, David Lloyd George, approached the chemist Chaim Weizmann who had developed a fermentation process which utilised the properties of bacteria to convert starch into acetone and butanol. This work was done for the company Strange and Graham Ltd who needed butanol for the manufacture of artificial rubber. Weizmann eventually found a bacterium, *Clostridium acetobutylicum*, which could convert appreciable amounts of the starch in maize, rice and horsechestnuts into acetone and butanol. He patented the process. The Weizmann process, as it became known, was of colossal economic importance in the subsequent build-up of a fermentation industry, an industry which yielded 'a wide range of other products, from vitamins to antibiotics, in huge tonnages throughout the world'.[5]

In the 1960s, a number of petrochemical companies turned their attention to using bacteria to convert the by-products of the oil refinery process into a cheap source of food. One promising idea was to grow bacteria or yeast on methane, dry the bacteria into a powder and sell it as animal feed or human food. The 'miracle' food which was going to solve the problem of food shortages was called Single Cell Protein (the name refers to the bacteria or yeast which are both single-cell organisms).

The only company to get Single Cell Protein off the ground was Britain's Imperial Chemical Industries, better known as ICI, founded in 1926, and 'imperial in name, and imperial in aspect', as an early planning document stressed. They called their product Pruteen.

In the fittingly entitled *Setting Genes to Work: The Industrial Era of Biotechnology*, journalist Stephanie Yanchinski relates how production of Pruteen entailed building a huge plant housing a giant fermenter to hold up to 150,000 litres of bacteria. The entire operation had to be kept sterile, a costly endeavour for such a large area, and involved designing new instruments and equipment. ICI filed over 100 patents for the designs, and ploughed £100 million into research and development.

When Pruteen was placed on the market as animal feed, it

flopped. Farmers did not buy it, and ICI failed to gain the economic edge with which it had hoped to get through the next years of research and development in other areas of biotechnology.

The production process forewarned of the safety problems of scaling up genetic engineering processes for the factory. In 1971, workers on the Pruteen project complained of discharges from their eyes, headaches, aching limbs, shivering and of tightness in the chest. Even after ICI installed better ventilation, 'there were subsequent incidents whenever the dust concentration of the dead, dried and fragmented particles became unusually high further down the processing line,' Yanchinski reported. 'From experiences in the enzyme industry, it was already well known that large quantities of dried proteins swirling on eddying currents of air presented a health hazard,' including that of allergic reactions, even if levels of exposure were extremely low.[6]

The same hazards to workers from such production processes remain as the industry expands with the addition of genetic engineering methods to the system; and as we will see in the next two sections, new hazards are becoming apparent, and do not just affect workers. Biotechnology is not the 'clean' industry it is often purported to be, as projected in images of a bright, gleaming, clean, efficient world of factory farms.

It has been interesting to observe how proponents of genetic engineering have put across mixed and even conflicting messages about it depending on the circumstances. On the one hand, it is called a revolution in the sense that it will dramatically change how we live, what we eat, who we are (biologically and socially) and so on: 'This is something different!' On the other hand, when that image doesn't work to the best advantage (for instance, when it suggests unknown consequences), we hear: 'This is really nothing different! Biotechnology is just like the things human beings have always done. And it is similar to what happens in nature. After all, genes mutate and recombine in nature.' The image veers away from factory floor with its hard, shiny stainless steel vats to a world of soft, green, environmentally friendly biotechnology. Pick which mythical image you prefer.

The problem is that both images paint a picture of biotechnology which does not suggest its scale, hazards, socio-economic effects, or the exceptional circumstances under which it operates. For the first time in human history species boundaries are being broken in ways never before possible. This is *different*: say, that human genes and cattle genes are purposely put into bacteria, two examples to which we now turn.

BST: A Case Study

Agriculture is feeling the first effects of genetic engineering, and one of the first of the new generation of agricultural products is BST or bovine somatotropin, a genetically engineered cattle growth hormone. Leaked documents from a meeting of the Eli Lilly company, best known for its pharmaceuticals, revealed industry's strategy for introducing genetically engineered agricultural products in Europe.[7] BST would be their test case for the entry of recombinant DNA technologies into agriculture; and Britain, their testing ground. The choice was strategic: Britain was considered the 'weak link' in the international chain, and it was assumed that its government would be more friendly than any other in Europe. It was true. Secret trials of BST were carried out in unnamed dairies with the blessings of the Minister of Agriculture, Fisheries and Food. Milk from BST-treated cows entered the milk pool unlabelled, that is, without any indication that it came from treated cows. This happened when there was no definitive evidence to ensure that milk produced by BST-treated cows is safe for human consumption, that BST injections are safe for cows, and that such use of BST is economically viable.

Attempts to authorise genetically engineered BST in North America and the European Community met with the protests of consumer groups, animal welfare activists, farmers' and women's organisations. Many organisations carried out information campaigns, consumer boycotts of BST milk, campaigns to ensure the mandatory labelling of BST milk products or to ban its use altogether. Many veterinarians and other analysts added their voices to the protests, publicly criticising the use of BST in milk production on the grounds of public health, animal welfare, environmental welfare, and its economic impact on dairy farmers both in the industrialised world and in the Third World.

There are many different kinds of hormones which carry out many different physical functions in animals and human beings. Bovine growth hormone (BGH) is a protein hormone which healthy cattle produce. As the name implies, it regulates growth and muscle development in the growing animal. In a mature cow, it also regulates milk production.

BST (bovine somatotropin) is another name for BGH. It is used as a veterinary drug which is injected into cows to increase their milk yields. In Europe, industry prefers to use the term BST rather than BGH, probably because it obscures the fact that the substance is a growth hormone. The use of growth hormones was discredited when a class of hormones known as steroids, used to stimulate growth in farm animals, was implicated as a cause of human cancer. In Italy, the steroid DES (diethylstilbestrol) administered to beef cattle became linked to infant cancers and adverse effects on the health of young children. The European Community eventually banned the use of synthetic hormones as growth promoters, but the ban does not include genetically engineered bovine growth hormones.

Before the advent of genetic engineering, bovine growth hormone, or BST, was never widely available. It was difficult and expensive to produce, and could only be extracted in minute quantities from the pituitary glands of the animals. That changed with genetic engineering. Now huge quantities of BST can be manufactured by placing cattle genes in bacteria.

The basic process is as follows. Genes are isolated from the cells of cattle, and processed in ways that enable them to be inserted into bacteria and work there. As the bacteria multiply very rapidly, so do the novel genes within them; the bacteria thus can produce large quantities of the bovine growth hormone. The BST is then purified, that is, separated from the bacteria and other substances in the soup necessary for the bacteria to flourish.

Four transnational drug companies are involved in BST synthesis: Monsanto, Eli Lilly (through its subsidiary Elanco), Cyanamid and Upjohn. These erstwhile competitors were now united regarding the introduction of BST to dairy farming. The 'other side' were the sceptical public, farmers and regulatory agencies.

BST has been used on cows in Britain, France, the Netherlands, Germany, the United States, India and the USSR. Dorothy Wade reported in 1988 that the four companies involved had already 'spent around $400 million between them to secure a share of a world market which could be worth $1 billion a year if BST receives clearance in both America and Europe'.[8]

To help achieve that clearance, secrecy and misinformation has been the order of the day. In the UK, the Ministry of Agriculture, Fisheries and Food officially sanctioned the secret BST trials. In the USA, the Food and Drug Administration was accused of ignoring its own regulations by the outspoken Samuel Epstein, professor of Environmental and Occupational Medicine at the School of Public Health at the University of Illinois Medical Center in Chicago and president of the Rachel Carson Council Inc. In addition, the company Monsanto, which was carrying out trials in the United States, was accused of suppressing unfavourable results by Professor David Kronfield of the University of Pennsylvania School of Veterinary Medicine in Philadelphia.

The Ministry of Agriculture, Fisheries and Food (MAFF) is responsible for the control of veterinary medicines in the UK. They insisted that BST was totally safe. When the independent London Food Commission carried out an investigation of BST, the MAFF told them that the trials were being kept secret for two reasons. The first was that the Medicines Act requires such trials to be confidential. The second was the need for the MAFF to preserve commercial secrecy. The MAFF is bound by law to be advised by its Veterinary Products Committee (VPC). Leaked information revealed that the VPC had not certified the safety of milk from BST-treated cows. In fact, the views of the VPC had been actively suppressed. Professor Richard Lacey of Leeds University complained that he was obliged as a member of the VPC to sign Britain's Official Secrets Act and thus was forced to remain silent about the known health dangers of these growth hormones. Lacey revealed that officials at the Ministry of Agriculture drew up a press statement ostensibly reporting on a meeting of the VPC two weeks before that meeting took place. 'It was very, very sinister,' he said, '. . . and it did not really take into account what had happened at the meeting.'[9]

The picture did not look much better in the United States. The two government bodies responsible for regulation of animal drugs are the Food and Drug Administration (FDA) and the US Department of Agriculture. Both side-stepped their own requirements for allowing the use of BST on animals and the marketing of milk from treated cows. The FDA exempted manufacturers from the usual regulatory procedures on the basis of allegedly confidential data. That exemption, Samuel Epstein said, is:

in apparent violation of the 1968 amendments to the Federal Food, Drug and Cosmetic Act, which mandate that the Agency must have a 'prescribed and approved' test method which the industry is required to provide, for determining whether the drug is being improperly used with resulting illegal residues in food.

In addition, the toxicological studies on which the FDA based its safety evaluation were:

irrelevant for the safety evaluation of biosynthetic milk hormones . . . In particular, the Agency has failed to require safety evaluation of milk from appropriate multi-lactational and multigenerational studies on a wide range of critical veterinary, let alone public health, concerns.[10]

The scandel forced the FDA to reassess its authorisation of BST, which it was doing at the time of my writing this book. The situation in the USA has caused the postponement of European authorisation until the FDA's decision is heard.

These circumstances, especially the controversies over secrecy and the safety to consumers of the food produced by novel methods, did not stop the UK's Minister of Agriculture John Gummer from granting approval, on 3 March 1990, to another experiment, the use of a genetically engineered yeast for use by British bakers. It was the first time the government had granted approval for the sale and consumption of food which had been altered by gene manipulation. The Ministry of Agriculture reported that genes from a related strain of yeast were added to the strain in question, and that the result was a yeast which would speed production of certain enzymes responsible for fermentation. The mutant yeast was authorised by two food advisory committees, but environmental and food pressure groups attacked the decision. Tim Lang of the London Food Commission responded: 'The government has barged ahead again in secret with an unproven technology for which there is no need or demand.' Again there was no legal requirement to label the new foods to inform consumers; and again the government failed to consider the consequences. (Yeast spores easily escape into the surrounding environment and, once airborne, it is possible that the yeast would contaminate the production process of other bakers and brewers.) A spokesperson for the environmental group Friends of the Earth said that we could only hope that a rigorous study of its environmental impact had been made prior to its approval.[11]

BST versus Animal, Human and Environmental Welfare

The available evidence regarding BST is so dramatic that it is worth a review.

The backers of BST first said that the genetically engineered version was exactly the same in chemical structure as the naturally occurring hormone. It could be injected into cows to increase milk yields 10–25 per cent, with no difference in the milk quality, with an increase in profit for the dairy farmer (her or his only cost the cost of the drug), an increase in food production, and with no harm to the animal. They said that increased hormone levels had not been found in the milk, and that the synthetic hormones are, in any case, safe for humans as they are not biologically active in the human body. Finally, BST would be especially welcome in Third World countries where poor people are starving.

All of the claims are either incorrect or questionable.

First, contrary to industry's claim, the chemical (molecular) structure of BST is not exactly the same as the naturally occurring hormone. In fact, BSTs differ one from the other. The series of manipulations involved in isolating the gene and inserting it into the bacteria result in slight alterations. Yes, the genetically engineered BSTs do have the same 191 amino acids (the building blocks of the hormone) as does the naturally occurring hormone. But

Monsanto's BST has one additional amino acid, while Elanco's has an extra eight. Also, synthetic BSTs may have a different three-dimensional structure and hence possibly different biological activities because they are made on bacterial ribosomes (cell components where proteins are assembled), instead of on cattle ribosomes. The differences may be slight, but they may cause adverse health problems.

Second, Epstein said that the 1988 Animal Health Institute promotional report which made many of the claims above omitted reference to a number of adverse effects reported in many of the trials, and also to data showing inconsistent yields in milk production trials. Positive claims were based on only a small number of cows. He listed a number of adverse effects on cows based on incidental findings in the few small-scale trials carried out, and potential adverse public health effects. Claims, for instance, that increased hormone levels are not found in the milk were suspect. Other studies from consumer-oriented and animal welfare groups concurred.[12]

Third, contrary to industry's claims, the question of changes in the composition and nutritional quality of milk is open. There is some evidence to suggest that the treated milk will have a different protein content and a higher fat content. A 1984 study of Holstein cows treated with BST showed a 27 per cent higher fat content in the milk, an effect reported in three other studies. Milk protein levels may rise, or they may fall, as a result of BST treatment; the levels of other nutritional elements may vary; and the levels of dead cells can be higher in treated milk.

Some reports added that if BST-treated cows require more concentrated feed, it is likely that there will be increased pesticide residues and fungal toxins (poisons) in the milk, contaminants which are known to be cancer-causing. Because BST-treated cows have an increased incidence of infectious disease, treated milk may have higher levels of viruses or other infectious agents; and higher levels of antibiotic levels may result from the increased incidence of infectious disease in treated cows. Misuse of BSTs, which many think is inevitable if licensing is granted, would increase the exposure of the general public to these highly potent biological agents.

Fourth, contrary to assurances from the industry that the cows do not suffer, there are a number of adverse veterinary effects. BST is unacceptable in animal welfare terms.

Cows treated with BST are hyperstimulated with regular injections – once every two weeks for instance – to keep milk production at a peak. BST is said to increase milk yields by 10–25 per cent, and sometimes up to 40 per cent. 'The effect of these growth hormones on the animal is to burn her out rapidly, so that within a few years she is exhausted from the speeding up of her metabolic processes,' reported H Patricia Hynes, director of the Institute on Women and Technology in the US. 'Cows treated with BST have more infections, particularly mastitis, an infection of the mammary glands . . . they suffer more from heat stress; and their fertility is reduced.'[13] Other adverse effects reported are an increase in the incidence of lameness due to the feeding of greater amounts of concentrated cereals to treated cows. There is worry over potential problems with stress-induced diseases, for example bovine leukaemia. Misuse of milk hormones has also been reported, both overdosing and illegal use of BSTs as growth promoters in calves and sheep. The Green Alternative Link (GRAEL, an association of European Green Parties) has warned that BST would be used to overstretch worn-out cows.

Ironically, genetically engineered BSTs are called veterinary medicines; the cows are not sick and the 'medicine' may actually make them sick. To paraphrase rural development activist Pat Mooney: the same companies who market BST will no doubt provide another product to soothe the ailment caused by the first one.

Speaking out against the use of BST at a conference in London in 1988, sponsored by the animal welfare organisation, the Athene Trust, German veterinarian Wolfgang Goldhorn said that BST is unnecessary and counterproductive, that it will lead to decreased resistance in cows to disease, and an increase in their suffering. Modern breeding methods have already resulted in cows needing more intensive care and service by farmers, their natural defence mechanisms having been reduced by such methods. BST aggravates all the problems of high-yielding cows: metabolic problems, reduced fertility, lowering of natural resistance to bacteria and viruses that are already in the cows' bodies and to salmonella and other micro-organisms with which they come into contact. Veterinarians, Goldhorn lamented, are becoming repair technicians, like car mechanics. And, of course, all of the problems the cows suffer can affect human health as well.[14]

Residues of the hormone *are* carried into the milk, and pasteurisation and digestion may create by-products which trigger an allergic reaction in some people. These effects could be especially significant for young babies, elderly people and pregnant women. BSTs may also stimulate the production of substances in the body called Insulin-like Growth Factors which are active in humans and pose a hazard to human health, especially in infants. Other types of growth hormone, which are produced by the effects of milk hormones in the cow, have been detected in the milk of BST-treated cows. 'The growth hormone is apparently tolerated by the human organism,' GRAEL at the European Parliament. 'It is, however, unclear whether hormone fragments will have side effects on the human body.'[15]

Finally, for people in the developing world, BST is useless and counterproductive. BST can only be used on high-bred varieties of cattle which need to be fed on expensive concentrates; hence it is an economic liability for farmers who cannot afford the extra inputs, let alone the purchase of the BST. Also, BST-treated cows need cooler environments because they overheat, thus making BST useless in hot countries.

The ecological principle is very simple. Only healthy animals can be expected to produce healthy food. Human health and welfare is intimately connected with the health of animals and with the rest of the environment. BST is but one example of the effects of hyper-industrialised agriculture. Another is the fact that the food chain – all aspects of agriculture and the world's food supply – is controlled by pharmaceutical and chemical giants like the oil company Shell, the drug company Monsanto, the chemical company ICI, a take-over which has been dubbed 'the birth of the genetic supply industry' by Pat Mooney.[16]

Effects of Industrial Animal Breeding

BST is just one aspect of high-technology breeding and maintenance of cows. The reproductive life of a high-technology cow is geared towards the breeding of a few types of superior cattle, by means of, for example, artificial insemination using sperm from prize bulls to create hundreds of thousands of calves. Prize semen samples can fetch thousands of US dollars, and, when frozen, can be shipped all over the world. Now cows can be subjected to other sorts of reproductive technology. Prize cows may first be stimulated ('superovulated') with hormones to produce many calves; if, say, twelve or more embryos are produced, they may be flushed out of the cow's womb. Some of those embryos can be frozen, while others can be implanted into the wombs of other 'surrogate mother' cows. In this way, one elite cow can produce larger numbers of genetically related offspring; and 'inferior' cows also can be put to use. Once embryos are removed from the cow, they might then be split into two genetically identical embryos.

All of these methods result in calves of similar or identical genetic make-up. From the industrial breeder's point of view, it makes genetically superior cattle; but from an ecological point of view, these practices are not beneficial. Genetic uniformity decreases the number of varieties – the 'genetic diversity' of agricultural animals. Crop varieties have suffered a similar fate. One consequence of genetic uniformity is an increase in disease vulnerability; another is extinction of certain varieties. In short, the general ecological principle is that the more variety there is among animals or crops, the better.

Critical analysts of the bio-revolution point out that the disastrous effects of industrialised agricultural methods are being repeated with new biotechnology, which more often than not has the short-sighted aim of 'efficient' conversion of animals into food or producers of food. As one vocal veterinarian in England, John Webster, summarised:

Enhancing the efficiency with which food is converted into flesh is now the approach most likely to attract the biotechnologist, who can manipulate a genotype by inserting novel genes, supplement the animal's natural hormones with genetically engineered hormones or accelerate genetic change by *in vitro* fertilisation and embryo transfer.[17]

As a European Community report summarised:

Modern technological agriculture is pressing for animals to be produced in line with industrial requirements . . . The use of reproductive, genetic and computer tech-nologies in modern medium-sized and large farms is designed to increase the productivity of the individual animals.[18]

The Economics of BST

Who profits most from BST? Industry promises that it will bring bigger profits to farmers (more milk = more profit), more efficient food production (more milk per cow) and a great boon to the Third World (more milk feeds more hungry people). But this arithmetic and logic does not hold up on further scrutiny.

The obvious economic insanity in using BST in the industrial world lies in the fact that in these countries cows already overproduce milk: there are milk *surpluses* in these countries. In Europe farmers are faced with milk quotas to limit production in order to keep prices stable. In 1985, surpluses of milk and butter amounted to 1.8 million tonnes.

As for the promise of enhanced profits, it leaves out of the equation farmers with small herds or those using organic methods who will probably be hurt. Farmers in Wisconsin in the US complained from the beginning that they never asked for BST, and that they would be run out of business. The Green Party alliance in Europe reported:

> The US Congress estimated that if BST were authorized in the USA, only farms with a herd of more than 50 could survive. One third of US dairy farms would have to give up production. On the basis of these calculations, not even half the dairy farms in Europe would have a chance of survival.[19]

Indian feminist and environmentalist Vandana Shiva has demonstrated the lunacy the use of BST promises to be for people in India. The already existing methods of breeding cows and producing long-lasting milk products in the rural regions of India are the most efficient and appropriate for the conditions there. By contrast, when intensive methods of dairy farming were exported from the industrial world, it produced 'junk milk' that goes off quickly under those conditions. In this case 'progress' brought failure and hardship. Such 'progressive' methods require developing countries to borrow money from the international banking system to buy expensive technology invented by corporations from other countries, thus increasing the dependence and debt of those developing countries.[20]

The BST example illustrates the dishonesty, the secrecy and the financial interests served by the bio-revolution and its new technologies. It raises so many political concerns: the environment, animal welfare, human welfare, development, Third World liberation. BST is not an isolated case but exemplary in its industrial approach to food production, an approach which contributes to the present double-edged food crisis: the production of too much and contaminated food in some parts of the world, and not enough in others. Factory farming of animals can run small farmers out of business, food can get adulterated, animals can suffer cruelly, and the soil be worn out whether or not BST is allowed in our industrialised agriculture. What it is important to recognise is that BST and other products of biotechnology also industrialise agriculture but are presented and sold as 'more natural'.

When BST became the subject of so much controversy and a glaring example of genetic engineering's problems – some old, some new, some unknown but becoming evident – many of those opposed to its licensing in Europe and the United States understood it in a political context, as a concrete example of the dangers implicit in genetic engineering. A few spokespersons in industry and government continued to insist that BST is perfectly safe, 'natural' and useful, but most commentators had to admit to the problems. Some thought it unfortunate that BST was the first genetically engineered agricultural product that had come on the market and given biotechnology a bad name. They admitted that BST is an example of dubious science and an unwanted product. Medical products, they suggest, are a better example of the effects of biotechnology, a better measure of its potential for success. But an examination of the case of the medical counterpart of BST, 'human insulin', presents many similarities. The case is, however, different in one interesting respect. The initiative to create genetically engineered human insulin was an early instance of the strengthening ties between academic research scientists seeking fame, and industrialists seeking fortune – or sometimes both seeking both – on the rising wave of biotechnology.

'Human Insulin'

Insulin is a protein hormone secreted in the pancreas of animals and human beings. It is indispensable for maintaining proper sugar levels in the blood. In 1922 insulin was isolated at the Toronto Medical School in Canada. Chemists at the pharmaceutical company Eli Lilly developed a viable system for extracting insulin from the pancreas of pigs and cattle and for producing it on an industrial scale. Eli Lilly have since dominated the world insulin market.

In 1982, genetically engineered insulin was marketed in Britain and the United States. Like BST, 'human insulin' is a protein product of genetically altered bacteria. It was called 'human insulin' because it was made from DNA which codes for human insulin, using genetically altered bacteria. Because it is 'human insulin', not pig or cow insulin, it was promised to be much better and safer for people with insulin-dependent diabetes. 'Human insulin' was going to be perfect. It wasn't perfect. The major predictions about its effectiveness in treating diabetes and the costs of producing it were wrong. In addition, the research that led to it raised questions about the safety and ethics of gene splicing and recombining.

Prematurely, enthusiasts argued that 'human insulin' would not inspire allergic reactions, as did insulin from foreign (animal) sources. Insulin from animals is slightly different in chemical structure to human insulin, and some people ωo have allergic reactions to it. This is patently true. The chemical structure of human insulin is known to consist of 51 amino acids or building blocks; pig insulin varies by one, cow insulin by three of these building blocks. There was also the worry of unknown long-term adverse effects from using the animal insulins, and whether they contributed to the circulatory problems and blindness which, among other health problems, afflict so many people who have had diabetes for many years. However, it seemed to be forgotten that the Danish company Novo Industries had long since developed a purification method for pig insulin which made it virtually (but not totally) non-allergenic. It was also known that a small number of people using the 'human insulin' had allergic responses to it. The purified insulin from pigs was as good as, if not better than, the genetically engineered version.

In addition, there were questions about the genetic engineering process. There are many unknowns in the process, and there are particular worries about contamination from the bacterial soup. Would pieces of strange DNA or other replicating particles used in genetic engineering methods prove harmful? Some studies showed an unexpectedly high antibody response to 'human insulin' which suggests that it is seen as a foreign substance by the body.[21]

Worse, by the late 1980s there was a growing concern over sudden deaths in young persons with diabetes who were switched from animal insulin to 'human insulin'. Reports had been coming to light that some users lost the warning signs of hypoglycaemia, the serious low blood sugar condition which requires immediate attention in people with diabetes. In the US warnings had been issued to users for some time, but British manufacturer Novo-Nordisk UK did not issue such warnings until 1989, by which time more than 200,000 people with diabetes were using the genetically engineered version.

One woman responded to the newspaper accounts:

Within the last few years, insulin-dependent diabetics

have been encouraged or, more to the point, told to change their insulin . . . to 'human' insulin [her quotation marks] . . . Since my doctor changed me to this new insulin a few years ago, I have experienced the same problems referred to in your news report.[22]

Many years before, when three different teams of scientists in the United States were competing to be the first to clone the human insulin gene, Harvard biologist Ruth Hubbard questioned the wisdom of the immediate research goal and the ultimate therapeutic goal of creating genetically engineered insulin. Hubbard, a contributer to many feminist collections on science and to *geneWATCH*, the publication of the Council for Responsible Genetics (formerly the Committee for Responsible Genetics), called it a high-technology 'fix' that did not take into account the whole complicated metabolic condition of diabetes. Insulin does not *cure* diabetes, she argued; creating huge quantities of it could result in an increase in dependency upon it, instead of a solution which looks at the condition in context, with attention to social and environmental factors such as diet, standard of living and life style. She added that if US industrialists were so concerned about offering people a cheaper alternative, as they kept insisting, then why not look into the economics of needles and syringes, which cost more per day than insulin?

When the reports of adverse effects and deaths associated with 'human insulin' came to public attention in Britain by 1989, it became clear that insulin therapy is neither medically nor scientifically straightforward, and that the marketing of 'human insulin' could be considered irresponsible.

Edwin Gale assessed the situation in the 25 November 1989 issue of the medical publication the *Lancet*. He argued that there was no reliable evidence to show that 'human insulin' was the cause of the increased level of hypoglycaemia and the deaths attributed to it, *but* he added that it is not a simple correlation to make. The cause might or might not be the 'human insulin'. But that is not the crucial issue. His two key points are that the problems attributed to 'human insulin' are major and unresolved problems with insulin therapy generally; and that there is no important clinical benefit in transferring patients from animal to 'human' insulin. 'Our real problem is the inadequacy of current insulin therapy, not the insulin species we use,' he concluded.[23]

We often take for granted that existing therapies of modern medicine are the only and the best approaches to dealing with an illness. It is difficult to question something like insulin therapy because it saves lives; because its remedy is easy to understand (it replaces something the body needs); because other approaches are not as medically certain in their effect. But no medical conventions are absolute; diabetes and its therapies cannot simply be reduced to the lack of insulin and prescription of insulin, both Hubbard and Gale implied.

Nicholas Russell of the Department of Life Sciences at Bromley College of Technology in the UK came to a similar conclusion about 'human insulin'. He reviewed the most recent scientific articles on diabetes. He concluded that a slow methodical research aimed at studying the pathways of insulin action in the cells was not as 'flashy' or as quick with a therapy as was 'human insulin'; but on the other hand, it would prove more useful and less troublesome in the long run. He reasoned that attempts to modify animal insulin products by genetic engineering methods 'may be more of a public relations coup than a genuine therapeutic improvement'.[24]

Considering Hubbard's advice at the start of the scientific adventure in 1976, and considering that medical and scientific opinion about insulin and diabetes does not universally support the development of 'human insulin', why did it happen? Because, as Stephen Hall observes in *Invisible Frontiers: The Race to Synthesize a Human Gene*, it was the biggest scientific competition of the century.[25]

Three high-powered laboratories in the US were involved in the race. Each wanted to be the first to re-create a human gene, put it into bacteria, and have it function therein. The race was not necessarily about creating a medically useful product. One of the laboratories began their work on a certain human gene, one which codes for a substance called globin. However, they then switched to the insulin gene because, as Hall relates, insulin was considered a 'sexier' protein: it carried emotional and economic weight as a medical 'breakthrough' and was attractive as a lucrative product for drug companies. In this way the insulin gene became the object of the race.

Each of the competing laboratories was backed by venture capital interests or an existing drug company. At Harvard University in Massachusetts, Walter Gilbert's team became involved in the newly formed biotechnology company, Biogen. At the University of California at San Francisco (UCSF), William Rutter and Howard Goodman's team were backed by Eli Lilly, world leaders in the sale of insulin. A third team came out of a collaborative effort between the City of Hope Hospital (Arthur Riggs and Keiichi Itakura's team) and another laboratory at UCSF (Herbert Boyer's team). They worked with the newly established biotechnology company Genentech, partly founded by Boyer and driven by the venture capital whizz kid Robert Swanson. This project was ripe for commercialisation. The wooing of the scientists and the machinations of the venture capital investors read like a soap opera. The competitive nature of the situation was perhaps an attraction (Boyer once compared his interest in science to football).

The genetic engineering experiments involved aroused public fears over the safety of the mutant micro-organisms. What if genetically engineered bacteria escaped from the confines of the laboratories? What if they found their way into the human gut? Hubbard and other concerned scientists joined members of the public protesting against this line of research at Harvard. The scientists promoting the research responded that potential hazards were under control. Moreover, they considered the threat of being stopped by public pressure groups as an affront to their rights to 'intellectual freedom'. Intellectual freedom was defined in this case as the protection of the interests of an academic elite, rather than the interests of ordinary members of the public who share that common democratic right.

The response of the academics reminded me of an incident when a group of women, including myself, who campaigned in York on reproductive rights issues, wrote to the vice chancellor of a university about our arguments with the human embryo research that was going on in the biology department there. Only feminists seemed to recognise that embryo research required the administration of superovulatory drugs to women and surgical interventions to remove the necessary eggs from women's ovaries. Ethical committees were equating egg removal with sperm donation, a procedure which merely requires a man to masturbate, and does not raise the kind of ethical and social questions unique to women's situation. We raised this oversight in our letter and were met with the brief reply that the university believes in academic freedom. We concluded that the freedom of scientists to pursue this line of research is invulnerable to the objections of the ordinary members of the public, we women, who are personally affected by it. There seemed to be no accountability to women (as members of the public) as long as the scientific establishment and reputable funding bodies safeguarded the research with the concept of academic freedom.

What about accountability in the case of the insulin gene race? Much of the experimentation was carried out in secret, since all the researchers were working towards a possible industrial application, towards the desired first patent on the process and product, and the dream of a Nobel Prize at the end of the line. When Walter Gilbert's team failed to get clearance in the US for experiments, the researchers went to Britain to carry them out in high-security laboratories which were formerly used for chemical and biological weapons research.

The Genentech team was credited with winning the race to find the human insulin gene. 'Human insulin' was marketed and prescribed. There were two other promises it failed to fulfil. Genetically engineered insulin was going to be cheaper than insulin from animals, and it was going to save everyone from a predicted shortage of pig and cow pancreas. However, within a few years it became clear that it was not cheaper and probably never will be. The original calculations that had pointed to a future shortage of pancreas were not accurate. But that was not really the major concern of industry, Stephen Hall concluded. The attraction for the companies involved was that the genetic engineering approach allowed them to control supplies at the source, a source which they had created, and for which now they had only to generate a demand – by weaning diabetics off animal insulin and on to the genetically engineered 'human' insulin.

The magazine *The Economist* reflected:

> The first bug-built drug for human use may turn out to be a commercial flop. But the way has now been cleared and remarkably quickly too for biotechnologists with interesting new products to clear the regulatory hurdle and run away with the prizes.[26]

The insulin example was going to be better than the BST débâcle. It was not, because the whole approach to the creation of utopian products is perverse. An ideal is held up: that this sort of science and technology is motivated essentially to serve human welfare. The reality is a union of powerful scientific and industrial interests which is motivated not simply by greed, but also by an industrial point of view: that of the pursuit of discoveries in highly specific ways which are removed from the context and subject of investigation. This kind of objectivity turns its subject – nature, animals, plants, human beings – into the objects of its pursuit of knowledge for prizes and for profits. Accordingly, the limitations of biotechnolgy are denied or not properly recognised. The only solutions are the ones the union of science and industry sells us. Other paths, other solutions, other kinds of knowledge, progress in terms of human welfare and environmental welfare suffer by comparison to the latest technological innovation.

Perhaps it would be helpful to consider something which we already understand to an extent, the environment-threatening phenomenon of global warming, which by most accounts has been caused by a century of modern industrial practices and continues to worsen because of present-day energy consumption. The use of energy by our human ancestors for the tens of thousands of years previous to our time never posed such a danger. By the next century, the earth's temperature is estimated to rise higher than it has been in the last 100,000 years.[27] Science and technology which continues along the same path, with the same mind-set, is likely to add more insults and more injuries to the tally. Genetic engineers may come up with a better product than either BST or 'human insulin', but that means little if it will create bigger problems than it might (or might not) solve. On the other hand, those paths which biotechnology closed down might be opened up to suggest other approaches to making sense of our world and to solving its problems.

3   Protection Against Pests

The responsibility to prove that it is risky should be
switched to a responsibility to prove that it is safe.
Paula Bradish, Women's Hearing on Genetic Engineering and Reproductive Technologies at the European
Parliament[1]

Why is there no life on Mars?
Because they had better genetic engineers than we do. (A
joke that was current when the UK's Royal Commission
on Environmental Pollution was investigating the safety
and risks of releasing genetically engineered living
things into the open environment.)

This chapter is devoted to the controversy over safety and
control of gene manipulation technology. It covers a great
deal of difficult ground: the first wave of the safety debate
and attempts at regulation of recombinant DNA work; the
second wave – why mutant organisms are being taken out
of the laboratory and into the field, and discussions of
several cases in which genetically modified organisms have
been released; and finally, a reassessment of the concept of
'risk assessment' on which claims of the safety of mutant
organisms are made.

Early Concerns and Self-regulation

By 1972 the discovery of restriction enzymes made it possible to manipulate pieces of DNA in test-tubes. In 1973 the first gene was cloned. In 1974 11 established molecular biologists called for a partial and temporary voluntary international moratorium on certain types of gene manipulation experiments after consideration of an experiment designed by the Nobel Laureate Paul Berg of Stanford University in California.

Berg wanted to study monkey tumour virus SV40 (Simian Virus 40) by cloning it in the bacterium *E. coli*. This would entail combining the genes of SV40 with the DNA of a bacteriophage (a type of virus which infects bacteria). The resulting virus could then be introduced into *E. coli*. As the bacteria multiplied, so would copies of SV40 genetic material within it. Berg's intended experiment highlighted one of several worries at that time. *E. coli* is a common bacterium normally found in the human gut. Altered strains might find their way into human intestines, thus introducing foreign DNA there. If the genetically engineered *E. coli* infected workers or escaped from the confines of the laboratory, it could spread disease among humans and the thirty other animal species in which the bacteria are normally found. The Berg experiment also raised a particular risk. SV40 was found to cause cancer in hamsters; the introduction of DNA from such tumour viruses or other animal viruses into bacteria, where cancer-inducing DNA would reproduce rapidly, posed a definite hazard.

Furthermore, because bacteria exchange genetic material in the wild, there is a fear that altered strains which escape could interfere with other populations of bacteria in unpredictable and possibly dangerous ways. There was a similar worry that, inadvertently, new pathogens (disease-causing agents) would be created through 'shotgunning', a process whereby the entire DNA of an organism is spliced into fragments for study. In *Genetic Engineering: Catastrophe or*

Utopia?, Wheale and McNally pointed out that 90 per cent of the DNA in higher organisms has not been characterised and may contain parasites, viruses and cancer-causing genes.

Berg was a member of the committee which called for the partial and voluntary moratorium on recombinant DNA work, a call which went out to the international scientific community in a letter published in three major English language science publications, *Science*, *Nature* and *Proceedings of the National Academy of Sciences*, in July 1974.[2]

Only a year later, in February 1975, a specially arranged meeting was held in Asilomar, California, to discuss the biohazards of recombinant DNA. Around 200 participants, mostly scientists but including legal experts, attended. The meeting was sponsored by the US National Academy of Sciences (an organisation to which the most acclaimed scientists in the country are awarded membership) and by two US governmental institutes, the National Institutes of Health (NIH) and the National Science Foundation (NSF). The moratorium, such that it was, was lifted. The majority of participants agreed that genetic manipulation would proceed, although it should proceed with caution. The majority view was also clearly against outside regulation of scientists. Guidelines were issued by the NIH, which is largely composed of scientists. Scientists would regulate scientists.

In order to proceed with caution (that is, to minimise the risk of escape of experimental materials), a two-tiered system of containment was conceived. The first tier was physical containment. All gene recombination experiments were to be done in medium- or high-risk containment laboratories, similar to those used when working with pathogenic micro-organisms such as the smallpox virus.

The second tier was a novel concept of 'biological containment'. Biological containment meant creating 'safety strains' of bacteria. Bacteria can be manipulated in such a way that they may live only under laboratory conditions. For instance, bacteria may be altered so that they cannot make the crucial chemicals necessary for survival. Only under artificial laboratory conditions would their chemical needs be met. If the bacteria escaped from those conditions, they would die.

Although biological containment is important, especially where physical containment has been shown to be unreliable, Wheale and McNally noted, 'Biological containment can never entirely eliminate risk to workers . . . genetically enfeebled strains may revert to robust forms.'[3] There are also incidental risks: for instance, the intestines of people on antibiotic treatment are easily infected by weak strains of *E. coli*. With regard to the safety of workers, therefore, there are serious problems with both biological and physical containment of organisms in large, scaled-up factory operations of production, recovery and downstream processing. (In addition, it seems to me that there is something seriously unsound about logic that suggests that the way to make biologically manipulated micro-organisms safe is by biologically manipulating them.)

Biological containment remains accepted today as a method to minimise the risks of deliberate releases. 'Debilitation mechanisms' are built into the engineered bacteria or virus, making it 'less fit' for survival in the open environment. For instance, organisms may be given a 'suicide gene', a DNA sequence which causes it to self-destruct once it has achieved its purpose, in the hope that this might lower the risk of the micro-organism interacting in unexpected ways with the environment.

From Regulation. . .

In the US, around the time of the Asilomar conference, the NIH issued guidelines which were mandatory for any scientist using public money for genetic engineering, but voluntary for industry and for privately funded researchers. The rules were implemented through local biosafety committees, which included representatives from public pressure groups, but which were dominated by representatives from companies and universities. One government body complained that these biosafety committees had 'too many members with backgrounds in genetic engineering and too few specialists in physical containment, epidemiology, ecology and large-scale fermentation techniques'.[4]

The partnerships between academic scientists and industry presented a conflict of interests, said Sheldon Krimsky, noted historian of gene manipulation technology:

> Most of the leading molecular geneticists, on whom we must depend for our understanding of the products generated by genetic engineering, have commercial affiliations. This may be a factor in how they assess the potential risks of genetically engineered life forms.[5]

Britain took a different approach to regulation. The Advisory Board to the Research Councils recommended in 1975 that the techniques should continue to be used because of their promised benefits, but under certain controls. A government report followed, and from that, the Genetic Manipulation Advisory Group (GMAG) was set up in 1976, with responsibility for advising on gene manipulation and for assessing risks and precautions on a case by case basis. The committee had a tripartite structure of five employer representatives, five employee representatives including trades unionists, and eight scientific specialists. Meanwhile, since worker safety was at that time the primary concern, protection was carried out through the Health and Safety Commission (HSC) which had been established in 1974 under the Health and Safety at Work Act. In 1978 the HSC issued genetic manipulation regulations which made it mandatory for anyone intending to carry out recombinant DNA work in the UK to notify both the Health and Safety Executive and the GMAG.[6]

In 1976, the European Molecular Biology Organisation recommended that anyone else contemplating recombinant DNA work should follow the UK or US guidelines. But these guidelines were later relaxed, by 1977 in the US and in 1984 in the UK. The scientific establishment pronounced that the dangers of recombinant DNA work had been overstated; nothing catastrophic had happened, and so gene splicing and recombination began to become an everyday event in laboratories around the world.

. . . To Deregulation

If in 1974 the consensus among scientists had been to proceed with caution, within a few years it changed considerably to one of 'no special hazards here'. In 1977, molecular biologists meeting in the United States came to a new consensus, that gene manipulation was far safer than they had originally supposed and that it posed no unique hazards. Wheale and McNally called it 'the false consensus' on two grounds. Firstly, the decision-making process was faulty; secondly, the scientific evidence which led to the conclusion was weak. As many other analysts also had stressed, to reach such a consensus, the larger questions of risk – social, ethical and moral considerations – were pushed aside. Thus discussion was restricted to certain technical aspects of risk which the scientists involved selectively placed on the agenda. One of the major reasons for the change in attitude was a scientific paper published in 1977 which showed that genetic recombinations happen in nature, in higher organisms; and since 'hybrid DNA' is produced in nature, it therefore was not particularly hazardous. This conclusion, Wheale and McNally respond, is not sound: the fact that gene recombination occurs in nature does not make gene recombination in the laboratory safe. None the less, the scientific lobby defused political opposition to recombinant DNA work in the United States, and generally diminished the role of the public in decision-making on the subject, at least for a time.

What about the claim that nothing terrible had yet happened as a result of recombinant work? One of the fears at the time of the Asilomar conference may have been borne out – namely, the risks to laboratory workers who, in the process of chopping up the DNA of bacteria and viruses, might be putting themselves in contact with cancer-causing substances. Five molecular biologists working with tumour viruses and cancer-causing genes (oncogenes) at the Pasteur Institute had contracted cancer, mostly bone cancer, at about the same time. The five worked in two adjacent laboratories. The probability that the cluster of cancers could occur without some common cause is remote. When the news was made public in 1986, two of the scientists had died, the others were seriously ill, and a sixth researcher using similar genetic engineering methods had contracted cancer. The French Social Security Agency recorded one of the deaths as occupational disease. After investigations showed that no fault could be placed on poor working conditions or research methods, it was considered likely that cancer-causing agents being used in gene manipulation work must have somehow infected the workers. 'It seems inevitable that E. coli designed to produce oncogene proteins will enter the gut of lab workers,' the magazine New Scientist reflected.[7] Critical analysts pointed out that there had been little organised risk assessment carried out on the hazards of the work. What I find even more disturbing is that this case, and the need for future risk assessments, has been all but forgotten. Genetic engineers often say that their early fears were unwarranted because nothing disastrous has happened yet. What then were the deaths at the Pasteur Institute?

Stephanie Yanchinski said in Setting Genes to Work that none of the cautionary steps taken by scientists or government regulators in the early years ever seriously restricted recombinant DNA research, except in Holland, and briefly in

Japan where all gene splicing experimentation was prohibited until 1981. Elsewhere scientists proceeded with gene manipulation work unhindered. Indeed they have been encouraged in this work: gene manipulation experiments have touched every single discipline of scientific research, and are considered indispensable for answering biological questions in physiology, biochemistry, biophysics and neurology, among other disciplines.

Why Release Mutant Organisms?

The question of the environmental risks posed by mutant organisms was revived in the 1980s. Researchers now wanted to take genetically manipulated organisms out of the confines of the laboratories and into the field, arguing that the practical rewards demanded it. Agricultural uses were among the earliest benefits cited. Proponents argue that manipulated micro-organisms, plants and animals may be used to improve crop yields, to facilitate growth, to protect against pests and other environmental stresses on crops.

Several approaches have been developed. In one, bacteria and viruses are turned into pesticides. The idea is based on the known intimacy between micro-organisms and plants: many bacteria and viruses live in relationship with roots, leaves and flowers of crop plants, and some of them act as pest deterrents in their natural state. Genetic engineering, it is argued, can turn organisms into more efficient, more lethal 'biocides', biologically based pesticides for killing crop and forest pests such as insects, fungi and weeds (plants growing where they are considered pests).

In another approach, genes may be introduced directly into a plant to add substances such as growth hormone and nutrients, or to build in disease resistance. Farm animals could also be altered to increase growth rate, to create leaner meat and to reduce their vulnerability to disease.

Biocides, the biologically based pesticides of biotechnology, are ripe for big business. Farmers spend around US$15 billion each year on anti-pest products, mostly chemical pesticides.[8] Chemical products have become increasingly unattractive due to the known health and environmental damage many of the varieties are known to cause, and there is a great deal of public resistance to the use of both chemical pesticides and herbicides (plant killers). Today, such chemicals take longer to be approved. Hence the interest in biocides which, it is argued, are a safe alternative to chemical pesticides.

However, many of the companies involved in developing biologically based agricultural inputs (seeds, pesticides, fertilisers) are in the chemical business. Some voiced interest in using both chemical and biological inputs together. The idea is to design crops containing genes which confer resistance to specific chemical herbicides. A farmer could plant the special designer seeds bought from the company, and spray the crop with the chemical herbicide bought from the same company. The crop, resistant to the chemical herbicide, would flourish while encroaching plants would be destroyed. These seed/herbicide packages could help a company maintain markets for its chemical herbicides, and also enlarge its markets with new products, the seeds which hopefully could be patent-protected. 15 major chemical companies were investing in herbicide-resistant crops by 1990. But the idea was so controversial that in 1988, Ciba-Geigy, the Swiss pharmaceutical company and now one of the agricultural biotechnology giants, decided against developing herbicide-resistant maize. (Maize forms the world's largest seed market.) A spokesperson for the company said that they had a certain sympathy with environmentalists who saw the idea as just another way to make money.[9]

Other beneficial uses of mutant organisms have been cited. For instance, genetically manipulated bacteria may be used for the disposal of biodegradable waste and for metal recovery from ores. The objective is to design micro-organisms which work more efficiently than the non-engineered versions already used in these capacities. Of medical interest are vaccines containing genetically altered live viruses, which when used on an animal or person constitute a release into the environment. The virus in such vaccines can be picked up by animals and people in contact with the vaccinated subject. (The same year Ciba-Geigy announced it would not develop herbicide-resistant maize, the company bought a 7.9 per cent stake of the US biotechnology company Chiron, which specialises in using recombinant DNA techniques to develop vaccines and diagnostics.)[10]

Warnings

The confident logic behind the use of mutant organisms, whether released in fields or confined to sheds, met with various counter-arguments. One argument identified the weakness of the 'efficiency of production' justification. Why try to create leaner, faster-growing livestock for meat eating in the name of efficiency of food production when we already know that it is more efficient to use land directly to grow plant food for human consumption than to use it for grazing or to grow animal feed?

Another counter-argument asked, as for engineering plants and animals against vulnerability to certain diseases, does it not seem likely that these mutants might prove more vulnerable to other diseases, considering that highly specialised breeding has increased disease vulnerability of plants and animals? (See the next chapter for a discussion of this.)

And another brought up the issue of unpredictable consequences. An early case which came to public attention was that of a farm animal, a pig, produced by scientists at the US Department of Agriculture at Beltsville, Maryland. The gene which codes for human growth hormone, plus a regulatory gene sequence (a stretch of DNA necessary to the working of the growth hormone gene), was inserted into a pig embryo. The resulting animal had the novel gene in every cell of its body, and it expressed high levels of growth hormone. The objective was to produce a faster-growing, leaner pig; however, the animal suffered such crippling arthritis and metabolic and heart disorders that it was killed to relieve it of its suffering. The sad irony is that genetic engineering has pretentiously been called an 'enabling technology'. In short, although molecular biologists and genetic engineers might be successful in one sense, gene manipulation introduces risk factors: geneticists themselves point out that they do not adequately understand the complex interactions of genes with each other and with other cell components.

Among these concerns, most prominent was that of environmental welfare: the risk of disruption of the eco-logical balance of an environment into which a mutant organism is released.

Environmental Risk

Any living thing introduced into a new environment may interfere with its ecology (defined as environment in relation to the living things which inhabit it). Some scientists say that the risk of releasing genetically altered bacteria, viruses and plants into an environment is the same as that of introducing any new organism. In fact, they say, the risk may be less. For instance, if just one gene is changed in an organism which already lives in an environment, say a soil bacterium, then it may cause less disruption than the introduction of an unmanipulated organism which is alien to that environment. They add that genetically manipulated organisms are monitored by scientists; while in the past, humans have moved thousands of species of plants, insects

and animals from one place to another, and these movements were not carefully monitored.

Other scientists, however, stress that the deletion of
even one gene in an organism could upset the workings of
all the other genes in that organism, and further, that
molecular biologists do not know enough about the interactions between individual organisms and the environment
to be able to calculate the risks involved. Ecologists con-
tended that if novel micro-organisms and plants were going
to be released, knowledge of the ecosystems involved was
needed.

Professor John Lawton, academic ecologist and a consultant to the UK's Royal Commission on Environmental
Pollution (RCEP), called the eagerness to release genetically
engineered organisms 'frontier science': 'forge ahead and
don't worry much about the consequences'.[11] He stressed
that the lack of knowledge about how genetically modified
organisms would act in the environment was equal to the
naive and dangerous lack of understanding about chemical
pesticides such as DDT in the 1940s and 1950s; these are
now known to persist in the environment for long periods
of time after initial use, and thus pollute air, water and soil,
and poison birds, insects, fish and people, destroying the
careful balance of both the crop cycle and the food chain.

To illustrate possible harmful outcomes, the RCEP used
the example of releasing a designer crop with built-in
resistance to a chemical herbicide. They estimated two
possible dangers and one possible beneficial effect.

First, the herbicide-resistant gene might spread, for
example in pollen, to weeds which would then become
resistant to the herbicide . . . Secondly, there is concern
that the engineering of plants resistant to herbicides
could lead to the greater use of herbicides, which could,
under some circumstances, be environmentally damaging.
On the other hand, the outcome might be environmentally advantageous if farmers were able to replace

environmentally harmful herbicide by a less harmful one or to control the weeds using lower quantities of herbicide.[12]

It is vital that, however low the probability of environmental catastrophe, the risks of releasing the products of gene manipulation into the environment are considered seriously.

When the go-ahead was given for several releases of genetically modified plants, bacteria and viruses into a number of test fields in Europe and North America, several public pressure groups, among them the Gen-Ethic Network in Berlin and the Genetics Forum in the UK, called for a moratorium on such releases, at least until the risks were adequately studied and better known. Distrust of the regulatory process was compounded by reports of several unauthorised releases in which researchers carried out field experiments in gross violation of existing regulations.

Unauthorised Release 1

The first known unauthorised release was a viral vaccine for farm animals. The Wistar Institute in Philadelphia and the Pan American Health Organisation carried out secret experiments in Argentina with a recombinant DNA vaccine against bovine rabies. The live-virus rabies vaccine, created by scientists at Wistar, was allegedly smuggled into Argentina in a diplomatic bag, where it was used on 20 cows at a research station in Azul; caretakers there came into contact with the cattle. The Argentine government had not been informed, and when they learned of it in September 1986, they stopped the experiment. Three of the caretakers plus some non-vaccinated cattle were tested to see if they had been affected. Blood tests showed that two of the men and all of the animal contacts had produced antibodies to the genetically engineered virus used for the vaccine. That is, they showed a response which suggested that they were infected with the virus. This is a potential risk with any live-virus vaccine, whether or not it is created with genetic engineering methods.

The genetically engineered virus allegedly passed from vaccinated animals to all of their animal contacts and to some of the humans involved in handling and milking the animals. How much further it spread is unknown. Milk from affected cows was drunk by farm hands and their families, and also went on sale.

Argentine scientists at the Centre for Animal Virology, furious that their country became the test site for experiments that were unacceptable in the countries in which the research on the vaccine had originated, pointed out the possibility of the engineered virus escaping and mixing with a 'wild' type of the virus which occurs in the area.

David Kingsbury, former director of a major US Navy biological warfare research laboratory, and chief architect of government guidelines on regulation of DNA technology and its commercialisation, was sympathetic to the releasers: 'Wistar must have felt that the regulatory structure was too stringent. We may be overregulating and pushing companies to test their products overseas.'[13]

Wheale and McNally urged a stringent regulatory framework controlling the introduction of new live viral vaccines to 'resist the well-established practice of exporting novel medicines and vaccines for clinical testing to countries where legislative obstacles appear to be fewer', and recommended that the scientists involved 'take on the responsibility to uphold the same uniformly high standards of safety throughout the world'. The vaccination of livestock, pets, sporting animals, wildlife, zoo animals and human beings with such vaccines has the potential to inflict harm on the subject vaccinated and on non-vaccinated subjects.

Live viral vaccines are living viruses which are capable of

mutation, possibly reverting to a virulent form, hence there is a grave risk that the use of live vaccines could instigate an epidemic of the very diseases they are designed to prevent. Even in the absence of mutation, live vaccines may possibly have some unknown long-term effects.[14]

The Ice-Minus File

Among the most notorious of planned releases was of a bacterium called ice-minus. Ice-minus is a mutant of the common bacterium *Pseudomonas syringae* which feeds off potato plants. Researchers found that naturally occurring *Pseudomonas* speeds the formation of frost on the leaves and flowers of plants. The bacteria produce a protein (a gene-product) on which ice crystals readily form, hence the idea for ice-minus. Ice-minus 'is designed to protect plants from frost damage', the science journal *Nature* explained.[15] But, of course, ice-minus, what it does and what it may do require a more complex explanation.

Ice-minus is a deletion mutant, a bacterium from which genetic material has been removed. Scientists isolated the gene which codes for the protein responsible for aiding the formation of ice crystals at higher temperatures than would occur if the protein were not present. The gene was called 'ice'. When the 'ice' gene is removed, so the bacterium's capability to aid ice formation should be gone. The idea was to spray ice-minus on crops and have them colonise the plants; once colonised with ice-minus, the plants would be protected from the naturally occurring bacteria. Ice-minus is classed as a pesticide because its purpose is to displace naturally occurring strains of *Pseudomonas* (the pest). The expected result is that the overall temperature at which frost damage occurs would be *lowered*; but presumably ice-minus will not *prevent* frost damage no matter what else happens in the environment.

Two groups of researchers were involved in developing ice-minus, one at the company Advance Genetic Systems (AGS) based in Oakland, California and the other under Steven Lindow at the University of California in Berkeley. Lindow received funding from AGS. AGS called the mutant bacteria Frostban and the press called it ice-minus.

When researchers were ready to field test ice-minus it became clear that although at least three different government agencies claimed and were competing for authority over regulation of released organisms, their attempts at doing so were causing more confusion than control; in addition, the implications of a release had not been thoroughly thought through by the agencies involved. The Recombinant DNA Advisory Committee of NIH granted permission for the release as early as November 1983, but the experiment was blocked by a district court order following a suit by the pressure group Foundation on Economic Trends, who used environmental protection legislation to halt the release. This was just the beginning of a series of events.

AGS was granted permission to field test ice-minus on strawberry plants in Monterey County, California in 1985 by another federal government body, the Environmental Protection Agency (EPA). The release was blocked by local residents and public officials dissatisfied with the secrecy surrounding the plans. The EPA subsequently revoked the permit when they learned that AGS had already illegally tested ice-minus in the open, on a rooftop garden of the company's building in Oakland. They fined AGS $20,000, later reducing it to $13,000. Still it was one of the largest fines ever imposed for a violation in agricultural research. The EPA reinstated approval for the release in September 1986.

Permission for testing was also granted on the state level, by the California Department of Food and Agriculture, on the basis of data suggesting that the bacteria would not spread and are not pathogenic to native plants and animals.

Environmental groups delayed the release. In April and December of 1987, attempts to carry out field trials in two different places were subverted by direct action: in the first instance, protesters uprooted about 80 per cent of the plants that were due to be sprayed on the night before the release was planned.

Meanwhile, AGS had threatened to take its business abroad if the EPA did not allow them to proceed. A news report stated that the company was negotiating with scientists and government agencies in Italy and Australia about doing the tests there. 'We won't give up,' one company official said.

The publicity over ice-minus attracted the attention of scientists throughout the US. Some offered their thoughts on possible risks. One was that ice-minus might alter the ecological balance of other bacteria in the environment; and the subsequent imbalance might affect plants and animals in the vicinity. Another suggested risk was that local weather patterns might be affected. Bacteria drift into the upper atmosphere, and *Pseudomonas* is thought to contribute to cloud formation. Ice-minus in the upper atmosphere might prevent formation of rain droplets, which could be a serious problem should more of these types of organisms be used on large tracts of land for farming purposes.

To judge the probability of ice-minus having an effect on weather patterns, the US Office of Technology Assessment commissioned studies which concluded that the likelihood was negligible. The UK's Royal Commission on Environmental Pollution duly noted it and warned, 'the example nevertheless emphasises the need for care to be taken about possible environmental consequences.' In the USA the Council for Responsible Genetics (formerly the Committee for Responsible Genetics) commented that although ice-minus was receiving a great deal of attention, it was an atypical case, and future cases would not be reviewed as carefully. One could not use ice-minus as an example of the ability of the regulatory system to handle future applications and releases.[16]

One researcher who failed to obtain approval from the EPA before he went ahead with a release experiment was Gary Strobel of Montana State University. Strobel added a gene to the bacterium *Pseudomonas syringae* in order to boost its production of a toxin which kills the fungus that causes a devastating disease in Dutch elm trees. Naturally occurring *Pseudomonas syringae* produce a toxin, but not enough of it to work dramatically. After injecting 14 trees with the mutant bacteria, Strobel injected the trees with Dutch elm disease fungus. If his bacterial fungicide worked, the trees would not be decimated. The incident raised the question of scientific accountability far beyond the risk of genetically modified organisms. The scientist was criticised for acting irresponsibly in injecting Dutch elm fungus into a region where it had not been present. This was 1987. It further came to light that, in 1983 and 1984, Strobel released genetically engineered *Rhizobia*, soil bacteria, at several locations without the proper authorisation.

The EPA decided that it did not have the authority to fine Strobel, who excused his behaviour by calling it an act of civil disobedience. Sheldon Krimsky of the Committee for Responsible Genetics admonished him, saying, 'You do not commit civil disobedience in the tradition of Gandhi and Martin Luther King by potentially placing society at risk.'[17]

Analysts of these events point out that what may start out as a harmless organism for pest control may turn out to be a pest itself. Although only a small percentage of invasions or introductions of alien species of plants, animals and micro-organisms become pests in a particular environment, the number of releases of genetically engineered organisms will increase enormously in the following years if releasers have their way.

Although most of these organisms will remain harmless, the sheer numbers involved will but ensure that a significant number achieve pest status and other non-pests may also have subtle effects in natural eco-systems.[18]

In 1990, another incident of unauthorised release was reported in Australia. Meat from 53 transgenic pigs, engineered to carry extra growth hormone genes, was sold to and eaten by members of the public who were unaware of the origin of the meat. The pigs, products of a joint venture between Adelaide University and Metro Meats, were transported to a slaughterhouse in breach of voluntary guidelines on releases of genetically engineered organisms.

Who are the real pests here? Perhaps they are the releasers. Perhaps what is needed is a Royal Commission for Protection from Genetic Engineers Who Want to Release Their Organisms. Or perhaps, just as many nations have imposed restrictions on the international movement of agricultural products and livestock to protect their countries from existing and new pests, there should be restrictions on the international movement of genetic engineers who travel to other countries to release their products.

The Institute of Virology Case

In Britain five authorised releases of genetically altered organisms took place by April 1988. They included mutant viruses, potato plants, and a bacterium. The first release, in the summer of 1986, was an altered baculovirus (a type of virus) which infects the caterpillar stage of certain moths. (Baculoviruses infect only insects and other arthropods, and cause disease in certain insect species). It was developed by the Natural Environment Research Council at their Institute of Virology in Oxford headed by David Bishop. The long-term goal of the work is to create a 'quick-acting viral insecticide', that is, a virus to kill a particular type of moth which damages pine trees.

The rationale is to improve on a known phenomenon: viruses found in certain parts of Britain can infect and kill caterpillars but do not harm people and animals. The Royal Commission on Environmental Pollution pointed out that baculoviruses have been used for decades as biological pest control agents and have a good safety record, but the naturally occurring viruses take several days to kill the pest; the viruses must go through several replication cycles before there is enough within the caterpillar host to kill it. The viruses act slowly compared with chemical pesticides, which act almost immediately. By inserting a gene into the virus which will make it produce caterpillar-killing toxin, the altered virus should act more quickly than the un-manipulated virus. The toxin gene will come from the bacterium *Bacillus thuringiensis*.

Speaking at the 'Action Alert' conference on genetic engineering in London in 1988, David Bishop explained that such artificial insecticides are necessary.[19] Why? Because monoculture crops (a crop with only one species of plant) are prone to infection; because people don't want to eat blemished fruit in our society; and because there are massive problems of infestation by insects in the Third World. Nothing was said of alternative solutions, such as changing agricultural practices to increase crop diversity, or changing our opinions about the appearance of the food we eat. Nothing was said of the possible negative ecological and economic effects as this new generation of agricultural inputs is placed on the world market (a subject discussed in the next chapter). Nothing was said about the presumption that a 'faster-acting' viral pesticide is desirable, especially considering the existing problems with the super-efficient, immediately effective chemical pesticides.

Bishop went on to explain that it would have been irresponsible to add the toxin gene to the baculovirus, and then release it without a thorough understanding of the possible consequences. Research at the Institute of Virology was therefore concerned with the assessment of risks associated with the introduction of a genetically engineered baculovirus into a natural habitat. Modifications were made to the virus in steps, not all at once, in order to minimise risks by ensuring that the uncertainties introduced at each new stage of development of new organisms would be limited to an acceptable degree.

The first field trial used a virus modified to contain a 'marker gene', a built-in 'tag' providing a means for the researchers to detect the viruses and monitor their survival. For the second and third releases, undertaken in 1987 and 1988, the virus was stripped of its outer coat (viruses consist of a core of genetic material surrounded by a protein coat) so that it was less able to survive in the environment. This is a 'debilitation mechanism', a form of biological containment discussed at the beginning of this chapter. The field tests determined that the virus could not persist in the environment, but that its ability to kill caterpillars remained effective for the one or two weeks that the virus survived.

The fourth release involved adding to the stripped virus a 'junk gene' whose expression (production of protein) could be monitored. The gene added was for a bacterial enzyme. The objective was to test whether stripping the virus affected the functioning of its genetic material. If the added 'junk gene' worked inside the virus, then it is likely that when the toxin gene were added, it too would function as a gene.

The Institute of Virology produced an educational video for public consumption entitled 'The Trojan Horse', which Bishop aired at the 'Action Alert' conference. The metaphor was spelled out by the narrator: 'Like a Trojan Horse it results in a faster conquest.' A viewer retorted, 'Like a Trojan Horse, you don't know the dangers it will bring.' Other scientist participants voiced worry over the removal of a virus's outer coat: 'When would it be put back on?' they asked, stressing that it was a highly risky business considering the lack of understanding of the ecology of microorganisms. 'You cannot engineer for both safety and efficiency,' another participant commented.

'The Trojan Horse' attempted to create an atmosphere to woo the viewer over to its message. Gentle (calming?) music and magnificent shots of beautiful English country-side formed the backdrop against which the narrator explained the project, how carefully researchers were proceeding, and how they were informing the public as they went along. This was true. David Bishop had taken great pains to talk to the public, to call for more public notices, to send out press releases, to ask for local government involvement and public discussion. He repeatedly stated that the researchers were cautious and systematic. Cautious and systematic but, as he admitted, not certain. At the end of that day, a woman spoke from the floor. 'I fear for my grandchildren. Why haven't all of you been more vociferous? What can be done about it? I am terrified by everything I have heard today.'

Just as in the ice-minus case, the Institute of Virology project was scrutinised in a way future cases will not be. Many of those following in Bishop's footsteps are not necessarily going to be as open. The Confederation of British Industry urged 'caution over allowing greater public access to information on genetic engineering develop-ments'. They criticised the plans to establish new regulatory bodies for genetic engineering, preferring that respons-ibility remain with the Health and Safety Executive. 'UK biotechnology is second only to that of the US ... there is too much at stake to risk the introduction of substandard [sic] regulations.'[20]

Secrecy, regulatory oversight, illegal experiments, legal experiments – these are part and parcel of the safety question. The danger is not simply the unethical behaviour of those releasers who bypass regulations, but also the narrowness of the definition of what constitutes a risk; who has the authority to decide how much risk is not too much risk; and equally, the attitude that there is 'an acceptable degree of uncertainty' which can be measured through 'risk assessment' while ignoring other political factors as if they are not part of the equation. To cite one such example, although it is easy to claim that releasing genetically engineered products will help developing countries, the tendency will be to reinforce patterns of debt and dependency of these countries, and to export unsuitable or dangerous products which cannot find a home in countries with more stringent guidelines on releases.

'Risk Assessment'

The recommendations made to genetic engineers about the process of releasing genetically modified organisms are conventional ones, and are none the less important for that: move slowly, be careful, don't get carried away, think about the best uses, think public safety, be open and beware conflict of interests. However, there is a fundamental dissatisfaction with this art of 'risk assessment', a process which has become central to the future of biotechnology.

Risk assessment is now seen as a form of scientific research designed to gain more knowledge about the ecological risks involved in the release of genetically manipulated organisms, and more ways of assessing and limiting those risks. If scientists, including ecologists, can assess the risk of a particular release; and if the probability of adverse effects comes within an acceptable level, then permission would be granted for that release – although the releaser must monitor it and have defined contingency plans in the event that problems arise. Risk assessment includes knowledge of factors such as state-of-the-art knowledge of the organism, how it was manipulated, where it can go and what it can do, its pathogenicity to other living things (if any), its new genetic structure, effects it might have on the environment, information on the nature, method and magnitude of the planned release, proximity to other living things and information on the target ecosystem.

Major responsibility is therefore put on releasers to show what the dangers and risks of their releases might be and how they may be averted or controlled.

Critics of releases make the point that the benefits of gene manipulations are not assured benefits and are therefore not worth the risks which the releasers assess. However, the process of risk assessment itself is also questionable. There are a multitude of *ifs* in the advocacy of, say, biocides. *If* genetic engineers could transfer the ability to eliminate pests to a biological source; *if* scientists could engineer the organisms, and *if* they worked exceptionally well, and *if* they could be cheaply produced, and *if* they proved environmentally friendly, and *if* they solved major agricultural problems, then would not everyone and everything benefit?

One problem is that risk assessment is in its infancy. Information on survival and persistence of micro-organisms is practically non-existent, according to John Beringer, a professor of microbiology and former chair of the Planned Release subcommittee of the Advisory Committee on Genetic Manipulation in Britain. 'We will only be able to approve release of things for which there's a gut feeling that that's all right.'[21] The ability to genetically manipulate organisms is outrunning ability to assess the risks of the experiments.

Second, the environmental issue has been wrongly reduced to a technical issue, divorced from social and economic factors, and it has become solely the concern of scientists. Although nobody assumes scientists to be infallible, not even themselves, there is a problem in the assumption that 'risk management can be improved by settling scientific facts before worrying about any social implications, or in assuming that scientists identify "real" risks, with additional public concerns being due to misinformation or irrationality'.[22]

A third problem is the cost of risk assessment, in terms of money and time, because of the pressure of time and money involved: risks are being underestimated in the scientific and economic races to push the technology forward. As a direct effect, by 1990 many of the new independent biotechnology companies which had sprung up in the hope of re-creating the fortunes of the computer industry were faltering. Viable products – the engineered life forms – were taking longer to develop and were more expensive than anticipated. In many countries, legal battles over patent rights, regulatory barriers and public protests were adding to the troubles of marketing biotechnology. This may prove its vulnerable point: public demands for safety may make it too expensive and thus too unattractive a business proposition. As one science policy analyst in Britain reflected on the future of biotechnology, 'The major constraint on the revolution is money.'[23] On the other hand, the failure of the smaller independent companies may put more of the industry in the hands of larger, multinational companies, who are in a much stronger position to push their products into the market place and fight legal battles over regulation and safety controls.

Beyond Risk Assessment

In H G Wells' *The War of the Worlds*, the Martians who invade Earth are finally destroyed, not by any means the earthlings use, but by an invisible, ubiquitous, modest, monumental force. Germs. The Goliaths from outer space are overthrown by terrestrial bacteria, strange and deadly to the Martians.

We fear annihilation, starvation, sickness. We fear nuclear war, the legacy of nuclear technology. We fear the end of nature, the legacy of chemical technology, chemical farming and industrialisation. Genetic engineering is presented to us as a force with the potential of eliminating some of our fears. An attractive promise, perhaps.

When confronted with the destructiveness of society,

and asked the reason why, German writer Christa Wolf replied with another question: 'Where did it all begin, men having to be heroes?'[24] I would say that her question is of utmost importance in understanding the direction of present-day science and technology, some of the 'whys' of genetic engineering, with its 'great discoveries', 'man creating life', its promise of power over living things, its definition of that power as the ability to destroy pests mightily, efficiently, instantaneously, completely – like dropping a bomb, obliterating 'the enemy'.

What vision is this of our day-to-day existence, always 'at war' with the creatures with which we share the earth? All this talk of engineering life has the whiff of death about it. Constructing viruses that kill more efficiently; creating bacteria that self-destruct with a 'suicide gene' or a so-called 'balanced lethal system'[25]; running the risk of ecological disaster. This is why the question of development and progress for many feminists has become: 'What are women's needs as an indicator of human needs?' in order to assess the full social impact – on women, children, men, peasant and tribal peoples who are being further pulled into the global economy – of biological engineering, the philosophy behind it, and the actual practices.

Have you ever seen a map of the world showing the geographical extent of contamination with oil from tanker spills? There is hardly a coastline unmarked with the thick black lines tracing the abused areas. I have hoped for the invention of a genetically engineered Super Bacterium which will chew up all that oil. I still hope. It must be a perfect Super Bacterium which only does good and no harm. It does not drift or mutate. It does not degrade any other molecule of oil except that which contaminates the waters, soils, marine life and birds. When its job is done, it then runs the oil spillers out of town so the damage never happens again. Then it goes away, back to the laboratory from where it came. At this point in my dream, I realise that the Benign Super Bacterium is about as attractive as the

Benign Dictator. It may work in the short term or for a specific problem, but other problems run deeper.

A lesson from the nuclear age is fitting.

It is not unreasonable to base an ethical judgment on the pros and cons of a range of foreseeable outcomes . . . The key issue here is not so much the magnitude of the risk (which in any case can never be accurately measured) but the scale of the potential threat . . . We still seem to have discarded the most fundamental ecological truth of all – that all parts of creation are interconnected in a complex web of life.[26]

4 Will Genetic Engineering Help Developing Countries?

If the social consequences of innovations are disastrous, the people whose lives have deteriorated or been ruined find little consolation in the fact that the innovators meant well or that solutions imposed upon them looked terrific on paper. Susan George.[1]

Biological engineering is being sold as a boost to Third World development. Gene and allied technologies will, it is claimed, undoubtedly contribute to the improvement of the world's food supply; they can be used for creating medicines, vaccines and diagnostic tests for parasitic and infectious diseases; and they will add to the repertoire of contraceptive methods. In this chapter I will examine the claims made in one of these areas: agricultural technology, which has dominated the discussion of biotechnology in the context of global development. This is not surprising, for agricultural technology has always been the main topic – along with population control – in the development discussion prior to the arrival of biotechnology. Human beings need food to survive. Where food comes from, who grows it, how, who consumes it, and who profits from it, are all significant factors for the food security debate. Farming practices, the role of specific food crops and access to farming resources fundamentally affect people's lives.

Before going further, one might well ask what is meant by Third World development. Susan George explains in *A Fate Worse Than Debt*:

> Beginning in the 1950s, a new breed of economists invented the notion of development – a word that has now become well-nigh meaningless . . . For many years, in many quarters (and still today), it was assumed that 'emerging nations' had a single goal and must follow a single path to reach it. Never mind the appalling poverty of imagination displayed: people in authority both in the North and in the newly decolonised countries wanted the South to become 'like' the rich, industrialised (often ex-colonial) powers.

This concept of development has become the model on which foreign aid is based. George criticises it as imitative, lacking roots in the local culture and environment of the countries involved, and in need of infusions of foreign capital, technology and ideas.

> It goes for growth, usually without asking, 'Growth of what? For whom?' Industrialisation is frequently its centrepiece, sometimes export agriculture relying on industrial inputs. The rich countries of the North nearly always built up their own industries on a strong agricultural base; the model conveniently forgets this and favours instant industrialisation over food security. Those who designed it were particularly scornful of small-scale peasant agriculture, the source of livelihood for most of the people in the countries concerned.[2]

Another, sometimes called alternative, concept of development sees the 'growth' goal as an obstacle in achieving real development; it starts from an appreciation of the diversity of needs of different people and their real conditions of living; it implies a change in the conditions of dependence on industrial countries. Organisations which espouse this concept of development talk in terms of equitable distribution of the earth's wealth and resources as a first step to achieving that change. It should be clear from the context whether the concept of development which I refer to is the 'top-down' development model which is measured in terms of growth, or alternative ideas of development based on the needs of real, living people.

The Agrarian Revolution

Genetic engineering is affecting the future of farming everywhere. The capability of putting foreign genes into plants is considered one of the most significant developments in agricultural technology, which encompasses plant breeding, hybrid seed production, farm mechanisation, chemical fertilisers and pesticides. Genetically engineered bio-pesticides and seeds are considered the most lucrative new products of agricultural biotechnology. Seeds are the most important agricultural resource of all.

Seeds are the first link in the food chain and are basic elements in plant breeding and agriculture. Seeds contain the genes, the embryo and the nutritive tissues which, when germinated, develop into plants. In his usual succinct manner, Pat Mooney, a Canadian engaged in international development work and with the Brussels-based International Coalition for Development Action, summarised: 'If a company controls seeds, it is well on its way to controlling its end product: food.'[3]

The marketing of seeds has always been high on the list of development concerns. The seeds industry is second only to the pharmaceutical industry in its monetary worth. The seeds market was formerly dominated by family-owned concerns until 1971 when a number of the large transnational companies, led by the likes of Royal Dutch Shell and Ciba-Geigy, began to acquire hundreds of seeds companies. By 1983, the two major agricultural inputs, seeds and pesticides, were firmly in the hands of one industry. Only 10 major companies controlled almost one third of the major cereal crop varieties. The seeds market was then worth about US$50 billion.[4]

In rich countries, we have thought of ourselves as the 'haves' and the so-called Third World as the 'have nots'. But it is the Third World which is blessed with most (95.7 per cent) of the earth's varieties of plants, insects, birds and animals. The industrialised countries of the North have been wholly dependent on the Third World for agricultural survival, that is for food. Most of the world's important agricultural plants originated in tropical regions, many of which are former European colonies. The agrarian revolution (as it is called) was initiated by plant collection from these places, plant exchanges between the 'Old World' and the 'New World', and preservation of plants and seeds in botanical gardens in the 'Old World'.

Past movement of plants from one country to another were often robberies by colonial representatives, either scientists, diplomats or travelling patriots. Crops which were removed by travellers, explorers and invaders became the foundation of Western agricultural supremacy. For example, new plant varieties such as wheat and maize were introduced into the New World (the Americas) by European expansionists in the fifteenth century. Military involvement, especially of the navy, aided plant collection, not only in suppressing revolts but also in helping with the collection and transport of plant materials.[5]

One of the most well-known plant smuggling operations was the transfer of rubber trees from Brazil to Southeast Asia in the nineteenth century by scientists of the Royal Botanic Gardens at Kew in England. Scientific study provided an excuse for such illegal transfers of plants from one country to another. The Brazilian economy was wrecked when the rubber monopoly was lost. Another example of the colonial and economic motivation of plant exchange was the removal of the breadfruit plant from Tahiti to the West Indies in the nineteenth century, in order to provide food for slaves on sugar cane plantations when sugar was an important luxury commodity in world trade.

The present state of the world's food supply is built upon grave injustices. Often the complicity of scientific explorers is glamorised. Scientists who now chronicle that history often play down the exploitation of indigenous peoples and resources, and celebrate the scientific explorer-hero who stalked 'unfriendly' jungles in search of scientifically interesting and useful plants, animals and insects. These were the plant hunters (often smugglers) 'who risked their health and lives in search of new crop candidates'. Similarly, such commentaries insist that botanical exploration today (gene hunting) is also for the general good, including that of the Third World.[6]

The industrial world today remains largely in control of the world's plant resources. The potential of new gene technologies revived industry's interest in plant collection; the objective is to gather species for their genes. Not only are plants being sought from 'exotic' locations all over the world (North and South, East and West), but also other organisms, including bacteria and fungi, are sought as a source of genes. The scientific interest in gene hunting today backs up the interests of industrialised nations and industries in gaining further access and control of the plant or gene resources. The new biotechnology fulfils an economic and a scientific agenda in a situation which is both new and old. Biotechnology is initiating changes in the organisation of food production; yet there is a continuity in the political conditions which brought us to this juncture.

The Genetic Supply Industry

Most of the world's plant resources are still found in the

developing world. Yet most of the scientific/industrial plant breeding is carried out in countries in Europe, North America and Australia using, for the most part, plant resources acquired elsewhere. These plant resources are today the 'raw materials' of an agricultural biotechnology industry in which they are held, studied and bred, turned into new varieties whose seeds are sold (at a profit) to farmers, sometimes farmers in the very countries where the original plants were found.

In 1983, the international development co-operation organisation, the Dag Hammarskjöld Foundation, devoted an issue of its journal *development dialogue* to 'The Law of the Seed', in which Pat Mooney referred to the take-over of the seeds and pesticides industries by a few large transnational companies as the 'birth of the Genetic Supply Industry'. High on its agenda was the desire to patent seeds. The Seeds Action Network, formed in 1985 to counter the genetic supply industry, believes that people are losing control over a fundamental resource for development due to the present surge in the privatisation of plant material by 'the seedsmen of the North' who are interested in exploiting new biotechnologies.[7]

The industrialised countries, not the Third World countries, are patenting plant varieties. Royal Dutch Shell, one of the big shots of the genetic supply industry, accrued 40 per cent of all the royalties earned on patented cereal seed in Britain in 1981. At the time, only three companies dominated the country's cereals market. If I bought a loaf of Hovis bread available at the corner shop, I would be buying from the Rank Hovis McDougall company, which had taken over at least 90 seeds-related enterprises.

Who is protected by plant patents? Certainly not farmers or consumers; certainly not the populations of the countries from which the raw material (plants) came from to make the novel varieties. Patents protect breeders and that means the new seeds owners, companies whose research and development programmes create new and lucrative varieties for sale and export. Patents on plants allow breeders a monopoly, thus controlling the seeds market.

A second problem facing developing countries is in the potential of biotechnology to change the modes of production of certain crops altogether. For example, if tropical plants may be engineered to grow anywhere in the world in laboratories or factories, or if they can be bred to live in colder climates, what happened in Brazil in the nineteenth century could be repeated many times over. Such research leans heavily on gene manipulation methods, mapping the genetic profile of crop plants, studying the relationships between DNA and characteristics, and growing crops in tissue cultures under laboratory conditions or by means of other novel methods, rather than in soil. For example, two of the most important agricultural commodities produced in the Third World, vanilla and cacao bean, are the subjects of such research. At stake is the financial solvency of a dozen nations. By the late 1980s at least two companies held patents on microbial processes for producing cacao butter. Similarly, alternative sweeteners or sugar substitutes have been researched, and marketed, for over a decade now.[8]

One might assess these new modes of farming certain crops as a positive change, in that it may leave developing countries in a position to grow food for themselves instead of being committed to growing and exporting luxury food commodities to the industrialised consumer countries. Aside from the hopes that such changes in production sites and methods would offer relief, Third World countries may not so much lose export markets with the loss of trade products, but they will certainly fail to benefit from such biotechnological innovations. In addition, as has been so often pointed out, the transference of production sites to 'bio-farms' would not be possible without using raw materials from the developing world, and the pioneering work of producers and scientists in those areas. Biotechnologists promise that these new systems of farming will be better for everyone, but ignore the complex social, political and economic factors which ensure a negative effect for some. Growing export food crops for people in these countries is not a matter of choice, but is a result of the international debt crisis.

Another worrying effect of the agrarian bio-revolution is the loss of agricultural resources, epitomised by the flow of genetic resources (plants, seeds and so on) out of developing countries and into the hands of the industrialised ones. The search by plant breeders for new genes for industrial use has expanded enormously during the past 50 years. Once called plant hunting, it is now called gene hunting. The patent-and-profit motivation is easy to grasp. What may not be so clear in this context is how industry, governments and science have a shared interest in collecting and controlling the stores of useful agricultural plants. Their aims and interests do differ, but they have an interdependent relationship in common, and the area where this becomes most evident is that of the control of seed stores, known as seed banks or gene banks.

Gene Banks

Seed banks are collections of plant varieties and are maintained by various research institutes, universities, international and national organisations, some of which are governmental and some non-governmental. In Britain, for instance, there are gene banks at Cambridge University, the Institute of Plant Breeding and the Royal Botanic Gardens at Kew, which has the most extensive collection of wild plant species in the world. Most of these collections emerged in the twentieth century for economic reasons and for farming-systems research. Some are former botanical gardens, but today they are known as seed banks or, more recently, gene banks.

In gene banks seeds are stored at reduced temperature

and moisture, and plant materials are stored in test-tubes or in field collections. The collections in gene banks can consist of the whole plant, the seeds, or the DNA of a plant species. Regardless of the form in which they are kept, even if they are collections consisting of whole plants, they may now be referred to as gene banks. The strength of gene-think is such that the material, concepts and language of genes displaces that of whole organisms.

Among the international seed storage centres, the most well known is the International Rice Research Institute (IRRI). Established in 1959, IRRI maintains collections of the rice varieties which still exist and it carries out research on them. IRRI was the first of 13 such research centres in Latin America, Africa and the Middle East founded to maintain seed collections used by breeding programmes worldwide. These centres are grouped within the Consultative Group on International Agricultural Research headquartered at the World Bank. The World Bank is an agency of the United Nations whose primary function is to make funds available to assist developing countries. However, the meaning of 'development' for the World Bank is specific to their objective to facilitate the growth of international trade. Its twin organisation, the International Monetary Fund (IMF), is the lender to which so many countries are in debt, and it imposes a similar economic orthodoxy of 'free trade' and commodity production on those countries, which are pulled into an international economic system whether or not they wanted to be there.

Hence, behind the stated purpose of tackling Third World agricultural needs is a political economic model which insists that development should not disturb business as usual. Vandana Shiva discusses the IRRI in her book *Staying Alive: Women, Ecology and Development*. IRRI was formed nine years after the establishment of a similar Indian institute, the Central Rice Research Institute (CRRI). CRRI was controlled by the Indian government and based its rice research on indigenous knowledge and indigenous plant resources. Its aims were clearly in conflict with those of US-controlled IRRI. The director of CRRI was removed when he resisted handing over CRRI's collection of rice seeds to IRRI, and when he resisted the introduction of a particular variety of rice from IRRI called a 'High Yielding Variety'. The choice of the power-brokers who supposedly were interested in Third World development was clearly *not* to develop seeds to withstand drought and pests in order to improve traditional methods independent of foreign supplies and procedures. What happened at CRRI was an instance of genetic imperialism: how control of farming practices is maintained by agricultural superpowers.

To understand the resistance of the CRRI director to the imposition of High Yielding Varieties, let us consider what they are. High Yielding Varieties, or HYVs as they are known in the business, are varieties of crop plants bred to have a built-in capability for yielding more food per seed. But to attain those high yields, HYVs require pesticides (which must be bought from the companies which manufacture them) and heavy irrigation (which requires a large, reliable water supply). The use of HYVs also requires an acceptance of the environmental and health damage that the use of pesticides causes. Furthermore, in certain regions where HYVs were initially successful, within a few years they failed to achieve high yields and were also found to be more susceptible to disease. In some areas, the land was spoiled after some years of HYV planting. In addition, the 'progressive' agricultural practice of introducing HYVs replaced thousands of native varieties with a few 'high-yielding' types. Variety was exchanged for uniformity, decreasing the wealth of crops available, and increasing the vulnerability to disease in those remaining.

HYVs were hailed as miracle seeds, the rewards of scientific innovation. The miracle seeds were not so miraculous after all. The progress promised (high yield) was too often short lived and costly. Some analysts refuse to call the seeds HYVs, renaming them 'High Response

Varieties' (HRVs) to reflect the requirements for large amounts of expensive agricultural inputs which are unaffordable for many developing regions of the world where agricultural investment and water are scarce. HRVs are wasteful of resources and risky; they have been known to lead to crop failure as a result of lowered drought and pest resistance, as occurred when rust and other diseases afflicted Latin America's entire coffee crop, which consisted of only one type of tree, the result of a practice known as monoculture.[9]

The Legacy of the 'Green Revolution'

HRVs were one of several elements brought to bear on the 'Green Revolution', the introduction of mechanised, pesticide-dependent agricultural practices introduced first in Mexico in 1943 and then in Asian countries in the early 1960s. (Here the connotation of the word 'green' is different to its present one: the 'green' in the Green Revolution referred to the hoped-for results of bumper harvests, not the farming practices which would hopefully result in them.) The other practices used in the Green Revolution were chemical pesticides, monoculture (the practice of planting one type of crop instead of many different varieties), and deforestation and destruction of wild habitats.

The food and environmental crises we face today are in part caused by these modern farming practices. The effects of health and environmental damage caused by many of the chemical pesticides and fertilisers are now well known. The practice of monoculture, in particular, has contributed to the environmental crisis in farming: the variety of crops the world uses for food has dropped from 3000 to a mere 20 species, resulting in the problem of genetic uniformity (lack of diversity of types of crops).

Crop uniformity is a result of changes in farming

practices over the past 100 years, beginning with mono-culture. An example of the devastation that can be caused by this practice is the Irish potato famine of the nineteenth century. The crop was planted from a few clones of a species of potato from the South American Andes. The entire crop was eventually wiped out by a fungal disease, leaving two million people dead. A similar process occurred with the maize failure in the US and the wheat failure in the USSR in the 1970s. In short, on a global scale, cultivation of a few, uniform varieties narrows the food base, increases disease vulnerability in plants, and leads to the extinction of many other varieties because they are no longer used. Monoculture diminishes food security. Gene-oriented answers to such problems are illustrated by the response to the maize disaster in the USA: 'genes from other varieties of corn were hunted out and used to rescue the American varieties', Omar Sattaur explained in the language of hunters and heroes.[10]

The Green Revolution has led to another similar crisis: to the extinction of species of plants, animals, insects and birds. Wildlife is disappearing at the astounding rate of 100 species a day; the main cause is the destruction of tropical rain forests in which half of the earth's ten million species live. The forests are being destroyed to make room for agricultural land for crops, often to grow fodder for livestock animals. Modern agricultural practices are largely to blame, Sattaur concluded. 'Our activities in the past century have done more to kill diversity than anything we have done in the 10,000 years since we began to domesticate food plants.'[11]

The Green Revolution aimed to bring modern farming practices and technology to developing countries. It depended on the initial participation of plant geneticists and the financial backing of the Rockefeller Foundation, the wealthy US organisation known for its ideological commitment to population control. Later, the Ford Foundation joined to finance early stages of the Green Revolution in

Asia. Prematurely, these foundations trumpeted their triumph in solving the food crisis. They were wrong: the food crisis continues in Latin America, the Indian sub-continent, Africa and Asia. The Green Revolution brought about land damage, destitution for many people, and further dependence on Western agricultural imports (fertilisers, herbicides, pesticides), and the need for more frequent irrigation.

As the Green Revolution continued, fertiliser prices quadrupled. It drew peasant Third World farming into the mainstream global capitalist economy, undermining sub-sistence agricultural economies. Whole ways of life were – are – being destroyed. Women's role as cultivators, their economic role and their status, was devalued as men were designated the recipients of the new technologies and became the cultivators and the farm machine operators. Susan George called the Green Revolution a 'horrible example of backfiring technology'.[12]

Was anyone aware of the consequences? Susan George argues that even agencies like the United Nations Food and Agricultural Organisation (FAO) which are supposed to act in the interests of the people in developing countries were aware. Technological utopias do not exist. 'Where nothing is done to alleviate inequalities, the Green Revolution is guaranteed to worsen them,' Susan George recognised.[13] The same holds true for the Gene Revolution.

The World Bank lent its support to the Green Revolution then; and it does so to the Gene Revolution now. It embraces the promises of a new kind of biotechnology as 'environmentally friendly' and productive for the de-veloping countries. Development aid in this sense has been and continues to be a means to promote industrial farming practices and the export interests of the seeds-men of the North. It takes many different forms: for instance, education exchanges which are not really about exchange (dialogue) but about the training of Third World agriculturalists along the scientific model of farming.[14]

Education exchanges may now be more appreciative of indigenous farming knowledge than they were in the Green Revolution, but the concept that now informs the exchanges is that which sees biotechnology as a panacea for all ills: developing countries need 'more genotypes' which gene technology can provide; meanwhile 'genetic resources' remain firmly in the hands of a few industries in the 'developed' world.

Controlling Genetic Resources

Third World governments have been largely dissatisfied with the control of crop resources by industrial countries, transnational companies and aid agencies such as the Ford Foundation. Official representatives from those countries have lobbied for binding international agreements for the free flow of plant resources. International networks of seed banks have been operating since 1974 to develop a world network for exchange of plant genetic resources, but the ideal of free exchange is far from met. For example, by 1970 the United States government had secured large stocks of wheat seed and crop freely gathered in Afghanistan and Libya; these varieties did not exist in national collections at the time. Mooney recounted in 1983 that: 'While the United States has obtained this useful breeding material free of charge, it is now applying political considerations to its exchange, even to the Third World countries that first made it available.'[15]

The first step in redressing the balance within the present scheme of things came in 1983 when the FAO formulated a non-binding International Undertaking on Plant Genetic Resources 'to promote the free sharing of plant genetic resources'. This first step followed a long hard struggle by the developing countries. The original proposal was put forward by Libya, Mexico and Peru at an FAO congress in 1981. At that meeting, delegates learned of a letter written in 1977 by T W Edminster, a high-ranking official of the US Department of Agriculture, to the chairperson of the International Board for Plant Genetic Resources (IBPGR), which had asked the United States to accept formal storage responsibility for a number of world crops. 'Edminster responded positively,' reported Mooney, 'but advised IBPGR that any material the US received on the Board's behalf "would become the property of the US government". While stating that it was the policy of the government to exchange material freely, Edminster felt obliged to add, "political considerations have at times dictated exclusion of a few countries".' Delegates dubbed the situation 'Genegate'.[16] Third World delegates to FAO conferences in the 1980s fought for a number of measures to regulate international conservation and free exchange of genetic resources. As Fowler, Lachkovics, Mooney and Shand reported in the 1988 issue of *development dialogue*, countries of the South envisaged an 'international gene bank' and an 'international convention' whereby any *bona fide* researcher or breeder could gain access to genetic materials, regardless of the individual's personal politics. All of these attempts met with opposition from industrial countries, and the goal of full and free exchange of all categories of 'genetic resources' – whether they be wild categories or patented varieties of northern corporations – continues to be controversial.[17]

The principle of free exchange of plants, seeds, and other natural resources challenges the concept of their ownership. The principle has emerged from discussions between many development organisations who say that the world's food resources should not be owned by any one individual or company or government but should be available for all people.

On the other hand, financial considerations aside, a promising by-product of biotechnology has been conservation work. Today, with one million species at risk of extinction by the end of the century, biotechnology –

especially gene hunting and banking – is being used to store varieties of crops which might otherwise be lost; to make them accessible to breeders in the future; to try to cope with impending environmental crises; to study the genetics of important food crops, to screen them; and to conserve endangered species. Animal species (livestock, *live stock*) are being conserved too, by the deep-freezing of the sperm, eggs and embryos of the animals. Some commentators have suggested that women's eggs, men's sperm, and human embryos could similarly be conserved in order to protect the species from genetic damage due to nuclear fallout, or other environmental dangers. (We would, of course, need other reproductive technologies in order to reproduce with these eggs, sperm and embryos.)

Such methods can serve environmental and ecological welfare, to an extent. However, there are problems. For instance, the European Parliament's Committee on Agriculture, Fisheries and Food cited an example of a possible negative environmental outcome from leaning on reproductive technologies for animal breeding. The survival of many species of orchid depends on the grazing habits of a threatened breed of cattle which lives on the slopes of the Black Forest in Germany.

> Native species must be kept in their natural habitat. As deep-frozen embryos or in sperm banks they cannot evolve and adapt, and similarly the ecosystem cannot survive without them as a necessary element.[18]

Saving and caring for dwindling species is laudable; so are attempts to rediscover old food crops which have been lost, sometimes because of violent suppression, as was amaranth by the Spanish conquistadors. But how conservation and rediscovery are achieved is another matter. To accept utopian biotechnological solutions to environmental degradation without further analysis, or without thinking about its risks, or about the most fundamental need of all – social change – is limiting at best, dangerous at worst.

Science and Accountability

Some might say that any attempt to halt the further extinction of species or solve the agricultural crisis will have problems. But I think that opinion ignores the divorce of 'political matters' from 'scientific matters' that takes place in advocacy of biotechnology. Scientists have shaped politics in the sense that the development and practice of a certain type of science is a political choice. Science is, however, often represented as apolitical: when science fails to deliver, we are told: 'More research (knowledge, technology) is needed!'; when the scientific agenda proves disastrous or worse, we hear: 'Politics got in the way.' Politics is always in the way.

Today, famine in parts of the world, in Ethopia and Bangladesh, for instance, is often blamed on 'natural disasters' such as drought or floods, and on lack of access to technology to counteract those natural forces. This neat explanation fails to research the history of the causes of particular incidents, which often show clearly that 'too much', not too little access to technology contributed to the 'natural disaster'. The catastrophic floods in Bangladesh in the late 1980s, for example, were caused by deforestation in the name of development, and not by a lack of access to technology.

Furthermore, it is no coincidence that the science of modern agriculture (high response varieties, chemical pest-icides, monoculture) just happened to serve corporate and political interests. Is it just a terrible coincidence that these practices caused environmental damage and ruined many people's lives? I think not. The science which predominates in our lives and which informed the Green Revolution has continually excused itself from questioning the reasons for the existence of inequalities and has, instead, often played a role in their perpetuation. Doubts should therefore be raised about the capability of the biotechnology of the Gene Revolution to redress the damage done to the environment, to biological diversity and to the food supply. Pat Mooney summarized, 'It is hardly prudent for the world to place its future food security in the hands of an uncertain science.'[19]

In her book *Staying Alive*, Vandana Shiva sees the present biotechnological 'imperative' as a continuation of mal-development practices, and she proposes a different model of development, one that integrates the ecologically sound practices of peasant women and tribal peoples in India, and one that reinstates the feminine principle which, she argues, has been expunged from the philosophy and practice of Western science since the European Age of Enlightenment.

In the Indian hill forests, women are the main growers of food, the sylviculturalists, as she calls them. If the forest does not survive, neither do they, their children and their people. She assesses their farming system as *ethno-scientific* and life-affirming, one that looks at the long term, not just short-term gains. She offers a positive lesson in the practice of productive forms of knowledge and action which oppress no human being, not women, not childrn, not men, not tribal or peasant peoples. The same cannot be said for the maldevelopment model which has devalued women's work and status, forcibly installed population control on women and men in Third World countries, and continues to threaten the very survival of whole groups of people and their environments.

What (mal)developers call 'wastelands' and wish to transform into privately owned revenue-generating land is common land which has fulfilled the basic needs of villagers and enabled them to practise and support their form of agricultural ecology. Women in India have been the greatest force in the resistance to the maldevelopers. A powerful image is contained in the Chipko movement against deforestation, where women literally embraced living trees in order to stop the destruction. Tree hugging proved successful in the face of the use of heavy machinery and police, and can perhaps be seen as a contemporary form of Luddism.

Shiva gives a deeply insightful analysis of power relations: the arrogance and folly of destroying functioning agricultural methods such as women's agricultural economy in India; the legacy of European colonial misadventure; the present lack of accountability not only of transnationals but also of the international agriculture research institutes in usurping control of resources (seeds and crops) and undermining successful indigenous agricultural methods, which, as Pat Mooney points out, have produced many of the most important genetic treasures, from ancient crops, the result of thousands of years of peasant farming.

The Population Factor

It would be irresponsible to address the politics of maldevelopment without addressing, however briefly, the theory and policies of 'population control' which have informed the current development discussion. It is difficult to ignore modern anxiety about increasing population numbers. In newspapers and magazines, in conversations throughout the world, population anxiety figures largely in discussions about the environment, food shortages, the energy crisis, poverty and education.

Population anxiety is underpinned by the belief that the world is suffering a terrible population 'explosion' that must be stopped, either because there is not enough food and energy to go round or because, even if we could manage to produce enough food, the environmental effects would be catastrophic. It is not uncommon to hear proposals for saving the environment from the stress of human needs placed upon it by implementing population control programmes.

The population 'explosion' and responsibility for its control are seen as the preserve of women, of poor people, of people who are uneducated, unhealthy or disabled, and the people of the 'underdeveloped' countries of the world. By contrast, overpopulation has not been considered such a problem for Europe, North America, Australia – at least not for white, middle-class groups (although even that is changing: individuals are often considered 'selfish' if they have 'too many' children).

Population anxiety conveniently avoids addressing the inequitable distribution of food, education, health and energy resources; it does not consider over-consumption and other causes of the energy, food and environmental crises; it does not consider the effects of different methods of farming, energy usage and consumption, and industrialisation.

The distribution of resources is just one example of what real population anxiety should consist. One quarter of the world's population consumes three quarters of its resources. Calculations made in 1987 showed that industrialised countries, accounting for 1.2 billion people, use 72 per cent of the world's primary energy sources (oil, gas, coal, nuclear and hydro-electric power). The developing countries, accounting for 3.8 billion people, use 28 per cent.[20]

Poet Adrienne Rich has written:

The decision to feed the world is the real decision. No revolution has chosen it. For that choice requires that women shall be free.[21]

The bio-revolution, with its sponsorship of the invention of further technical solutions to population 'anxiety' – that is the control of certain populations – by such means as the recently developed genetically engineered anti-pregnancy vaccines, does not make this choice.

Later in this book, I address some other aspects of biotechnology's impact on reproduction – immunological control of fertility through anti-pregnancy vaccines, and some theories of eugenics and practices regarding 'quality' reproduction. To summarise the above discussion on the way in which population anxiety provides yet another rationale for developing new biotechnologies, I would emphasise that targeting population as a major global development problem is wrong, that is, the wrong target is being 'attacked'. We know that in regions where policies have been implemented to empower women and thus raise women's status – for example, through providing education or employment opportunities – population growth decreases. As health care and other conditions improve for women and children as well as for men, then population growth decreases. No one factor taken out of context is going to provide a humane answer to human needs. In addition, to take 'population' as the problem, as if we all consume the earth's resources equally, is dishonest and it is built on a history of racial and sexual oppression from which it has not escaped.

In short, the question of the use of biotechnology for development is not simply one of who controls the technology used, but one of what kind of technology, and what kind of control. The energy crisis is a good example. If energy consumption continues at its present rate, even if that consumption were evenly distributed among the world's human population, the environment would still be overstressed.[22] Overconsumption of natural resources is itself a problem. In challenging it one challenges the economic concept of infinite growth and the model of development. The countries, corporations, institutions and we individuals who are overconsuming must change.

This chapter has been concerned with exploring one aspect of biotechnology's impact on food security, namely, the control and use of plant genetic resources. There are other aspects of biotechnology which are important to food production as well, such as biotechnological developments in the food processing industry, biotechnological research on artificial flavours, and the shift in focus to agricultural inputs and processing rather than on high yields (the focus of the Green Revolution). Also, the impact of biotechnology on the food and health industries must be looked at together to make a true assessment of biotechnology's worth to Third World development. All of these aspects together make up the Genetic Supply Industry referred to at the beginning of this chapter. Chemical, petrochemical and pharmaceutical concerns, as well as the giant food processor companies of the world, are in this respect one industry. Under such circumstances it is the poor and the powerless who will feel the squeeze. So how can biotechnology play a positive role in development?

I do not think it is particularly helpful to say that biotechnology can play a positive role under the right circumstances, because the circumstances have shaped what biotechnology is. As I pointed out in previous chapters, the pursuit of profit is a fundamental shaper, but it is not the only one. Much science carried out in academic laboratories with no direct links to industry remains a problem. Shiva and others talk in terms of an ecological approach to food production which takes account of nature and needs together. I think that a biotechnology which takes such an approach would look very different from how it does today; so different it would not be recognisable to us as biotechnology. It would be very different from the biotechnology which serves the Genetic Supply Industry.

5   Who Owns Life?
The Patenting of Living Things

A subject matter of an invention shall not be considered unpatentable for the reason only that it is composed of living matter.
Article 2 of the European Directive on the legal protection of biotechnological inventions[1]

Genetic engineering changes the hereditary elements of living things. Its most trumpeted outcome is the creation of mutants: mutant genes, cells, bacteria, viruses, flies, sheep, mice, potatoes, petunias and ultimately perhaps people, too. Mutants can be created by adding or subtracting or rearranging genes. If the mutant carries genes from a different species, it is called *transgenic*.

The first transgenic mammal to make the headlines was a mouse carrying a rat gene for growth hormone. Rat DNA was injected into the fertilised eggs of mice. Surrogate mother mice bore the progeny. The mice born (only seven transgenics were produced from 170 treated eggs) were bigger and carried the rat gene in every cell of their bodies. The first of them indeed grew to one and a half times the size of a normal mouse. They also suffered adverse physiological effects including infertility. Researchers went back to the drawing board.

Countless transgenic living things have been created. Sometimes the changes made are simply a matter of scientific interest, but examples are always being given of beneficial uses of adding and subtracting genes to and from plants and animals – such as the creation of drug-producing bacteria and sheep, (sheep which carry the human gene for Factor IX, a clotting factor need by some people with haemophilia, have been created using gene transfer technology in conjunction with embryo manipulation technology), pest-killing viruses, frost-preventing bacteria, leaner pigs, disease-resistant chickens and tobacco plants, pest-resistant potatoes, fast-growing fish. The creation of transgenic human beings is not considered in the same way, although the theoretical possibilities for social engineering purposes have been discussed by ethicists and philosophers, among others. Genetic engineering of human cells for therapeutic reasons definitely is on the agenda, however, but how important the genes of other species might be to the process is debatable.

Another kind of mutant predates direct gene manipulation (that is, gene splicing and recombination). Produced by combining cells from two or more different sources, the resulting creature is called a *chimera* or monster. In Greek mythology, a chimera was a fire-breathing beast with a lion's head, a goat's body and a serpent's tail, which gave us one meaning of the word, 'an idle or wild fancy'. Biology took the word and used it for a composite animal exhibiting two or more different genotypes derived from two or more different embryos which have fused at an early stage to become a single embryo. Some of the chimera's tissue carries one set of genes from the one cell source, while other tissues carry a completely different set of genes from the second cell source. It is known to happen in nature, but experimental chimeras have also been created by a few researchers to study development in mammal embryos, although the level of mixing of species was limited. In 1982 scientific history was made when a sheep/goat chimera in the UK was born – a 'geep' – created from methods of micro-manipulation of embryos developed at the Animal Research Station, Cambridge, formerly a government-funded laboratory, now Animal Biotechnology Cambridge Ltd, a private company which boasts of itself as 'the first company to offer a transgenic service'.[2]

Cloning is yet another sort of genetic engineering. Genes are not mixed up, as in transgenic and chimeric breeding. Cloning is the creation of genetically identical genes, cells, plants, or animals. In the realm of animal engineering, this includes creating genetically identical herds of cattle or flocks of sheep, examples of new, highly productive animals for farming, as the director of the Institute of Animal Physiology in Edinburgh has stated.

One of the first scientists to accomplish five clones from one sheep embryo was Dr Steen Willadsen, a former Cambridge scientist. He did this by removing the nucleus of a sheep embryo cell (the nucleus contains the genetic material); then he fused the empty cell with another complete cell taken from a second sheep embryo. The second sheep embryo already had developed to the 16-cell stage, a point at which the removal of one of its cells does not destroy it. Both the fused-cell embryo and the second embryo were placed into the wombs of female sheep. Both were born, and proved to be genetically identical.[3]

This realm of clones, transgenics and mutants lends itself to images of Frankenstein monsters created by a science that is out of control. *Frankenstein: or the Modern Prometheus* was written by Mary Shelley in 1816 while touring the Alps with Byron and Shelley. Byron suggested that they should each write a ghost story. Shelley's contribution portrayed the fictitious Dr Frankenstein, a gentleman scientist, who represents a new breed of European man, full of the spirit of the Scientific Enlightenment and Rationalism. Dr Frankenstein is obsessed with experimenting with electrical forces which at that time were thought to impart life (a theory which interested Mary Shelley's friends). Dr

Frankenstein constructs an apparatus to electrify a stolen cadaver. (The theft of bodies for use by gentlemen interested in dissection and exploring physiology was common at that time.) Dr Frankenstein's creation, known to us as the monster, awakens to new life. The doctor cries, 'I, Victor Frankenstein, had achieved what many men had tried and failed. I had created life!' The culmination of his quest, however, is the ruination of the lives of everyone around him, including his own and that of the poor monster.

Hollywood has interpreted Dr Frankenstein as the mad scientist hidden behind the gentleman's cloak; an unfair reflection of Mary Shelley's treatment. Other interpretations deduce from it that there are boundaries to knowledge, that humans should not go too far, and there are some areas we should never explore or know. I think the story is not so much about where to stop looking for answers to questions about the world; I think it is about what kind of questions are asked and how they are answered.

The modern (or should I say postmodern?) lesson of Frankenstein is: it's not what you know, it's how you know it.

If Dr Frankenstein were at work today, he would probably have taken a patent out on his monster. Patenting life has become the most contentious of the many contentious issues surrounding genetic engineering. The question has become, 'Should we or should we not allow the patenting of living things?' The question itself says more about biotechnology than does any answer.

Patenting Life

Shall all life be reduced to the status of a commodity to be sold for profit? Are plants, animals and people ever the private property of researchers and industries? Apparently the answer is yes, in genetic engineering. Laws on patenting, intellectual property and copyright are being changed to accommodate the ownership of genes, cells, plants, animals, parts of plants and animals, and human body cells and human genes as well.

Changes in the patent system in the last decade to accommodate 'biotechnological inventions' have been strongly contested nationally and internationally, for developmental, environmental, spiritual, animal welfare, religious, humanitarian and feminist reasons. The arguments against patenting are many, and include all of the concerns which we have investigated so far in this book. Broadly, they are:

- Patenting is morally repugnant: it reflects a human relationship with other living things that has been exploitative, takes it further and makes it worse
- Economically patenting hurts the most vulnerable: for instance, in agriculture farmers will be pressed to pay royalties on seeds, which will strain development in 'Third World' countries and place more control of resources in the hands of industrialists and powerbrokers
- Patenting encourages the creation of more genetically modified life forms, and their sale as profitable products; the releases necessary to test such products pose risks to the environment and human health; geneoriented breeding discourages biological diversity
- There are profound, though less tangible, effects. The definition of living things is being changed to accommodate technological systems and patenting – living things are often called 'living matter'. Patenting of life degrades who we are and what living nature will be.
- Patents in biotechnology do not even succeed in delivering the increase in creativity and innovations which may serve society as they promise; rather, they limit attention to one kind of product – a patentable kind, such as a novel biological pesticide – and waste

resources: for example, money which could be spent on research and development is spent on legal wrangles and patent disputes. Speaking about the Dutch experience with exclusive monopoly Plant Breeders Rights, an official of the International Federation of the Seed Trades commented in 1976 that the number of firms engaged in the cereal seed trade had decreased drastically because of it. The claim that PBR, a patent-like protection, stimulates the involvement of more companies was, in their experience, wrong.[4]

The idea of patenting living things is not totally new. In the last 100 years there have been numerous disputes over whether or not plants should be patentable. For instance, in 1877 the German Federation considered plant patents within industrial property legislation; while in the 1830s Pope Gregor IV attempted to secure property rights over all plants on all Vatican property in the various Italian states.[5] Moral taboos on patenting life have always played a role in these disputes, but mostly political and economic factors shaped decisions made for or against patenting. Political and economic factors do often affect moral norms, but the effect is particularly striking in the world of patents and proprietary rights.

What is a Patent?

Ancient Greek records of monopoly rights can be found as early as 200 BC. Patenting legislation arose as early as the fifteenth century in Venice, and since then, patenting laws have been altered in consideration of changes in the economic climate. The patenting of living things did not seem so necessary until the development of breeding into a technologically based industry.

By definition, a patent confers exclusive monopoly rights to an inventor to exploit a product or process for a limited period, usually 17 years. Patents are granted in exchange for the disclosure of an invention. One argument in favour of the existence of the system is that society benefits from the use of the inventions, a benefit measured mostly in terms of economic growth. It is also argued that patents, by rewarding the inventor, stimulate research and foster technical and scientific progress (research can be and is stimulated in other ways as well).

Patent law takes care of disagreement regarding ownership, where ownership equals the right to control the item, even to destroy it if the inventor so chooses. Black's Law Dictionary informs us, 'The essence of ownership is the right to control.'[6]

Patents can be granted for processes, products, and for specific uses of products. The three basic criteria for granting a patent on an invention are that it must be *new* (it does not exist already); *not obvious* (it involves an inventive step); and *useful* (it has an industrial application). The criteria for granting protection (not necessarily a patent) on plants are that the new variety must be *distinct, homogenous* and *stable*.

Criteria such as these prove difficult to assess when applied to patenting life forms. Genes, cells, animals and plants are not inventions; they already exist and are only being modified. 'Who will have the guts to declare a gene novel and non-obvious?' one critic asked.[7] Living objects are also not stable; they are changing all the time the way living things are wont to do.

Several analysts of the impact of the Gene Revolution on Third World development, among them Calestous Juma in *The Gene Hunters* and Pat Mooney in 'The Law of the Seed', point out that the concept of patenting was made legitimate at the International Convention for the Protection of Industrial Property held in Paris in 1883, the first international agreement on intellectual property rights. Around the same time, the Berlin Convention of 1884–5 legitimised the claims of various European powers to colonial territories.

These two developments, and the contemporaneous growth of the international division of labour, consolidated the European powers. The Paris Convention has been revised six times since 1883, each revision extending the exclusive monopoly powers of patent holders, and encroaching on those of ordinary people.

Plant Patents

At the turn of the twentieth century, scientists rediscovered the work of the nineteenth-century monk, Gregor Mendel, whose experiments with pea plants led to the classic 'laws of heredity' and the science of genetics. The rediscovery also sparked off a new era of plant breeding, a coincidental rise in the economic role of plant breeding and a renewed enthusiasm for plant proprietary rights, first felt among the breeders of ornamentals, especially roses. Plant breeders have since secured various forms of protection, but true patenting was not one of them.

The first living things allowed to be patented to any degree were plants, but only for a specific classification of plants called 'varieties'. A 'variety' is a category below the species level. A *plant variety* consists of a group of individuals that differ from other varieties of the same species, but can interbreed with them.

In the United States, the Plant Patent Act of 1930 was the first to allow the granting of patents on non-sexually reproduced 'varieties'. (Non-sexual reproduction entails cutting or grafting to produce a plant hybrid, and thus requires a degree of intervention that does not occur in nature.)

In Europe, instead of patents, a system of Plant Breeders' Rights (PBR) was set up in 1961 at the convention of the Union for the Protection of New Varieties of Plants. PBR grants the breeder control over marketing and sale of new varieties, but it does not protect the techniques developed and used to produce them. PBR guarantees that any breeder can freely use new methods to improve crop and animal varieties, and that farmers maintain the right to re-use seed from the harvest. PBR attempts to avoid a complete monopoly which would prohibit breeders and farmers from using seeds and processes under patent protection: total patent protection of plant varieties might slow down the development of new varieties and their commercial use in agriculture and reduce the profit margin. PBR also strengthened the ties between breeders and the seeds trade.

Economic times have changed. With gene and cell technologies at their disposal, not to mention a monopoly in the seeds market, industrialists and many researchers now wish to patent plants in the fullest sense of the term.

The pro-patenting European Commission tried verbal gymnastics to overcome the already existing obstacles. PBR already protects plant varieties, and its rules say protection cannot be duplicated by any other means. In consideration of this obstacle, the Commission decided to ignore 'varieties'. They said that patenting of plants should be allowed for all biological classifications *except* varieties. They added that any patent protection of plants or animals extends to the progeny. For seed crops, this means farmers would no longer be allowed to use seed from a previous harvest if those plants (or the original seed) were covered by patent; a user would not only pay royalties on the novel organism or seed, but on the following generations for 17 years (or for however long the patent protection is granted). When this set-up angered North American farmers, who joined themselves to other critics of patenting, the American Biotechnology Industrial Association shrewdly broke up the coalition by promising to exempt farmers from having to pay royalties on next-generation seed.[8]

The European Patent Convention of 1973 explicitly excludes patenting of plants and animals. However, generally, the boundary between patentable material and unpatentable material has always been in dispute. During the 1980s, the situation became more chaotic and confusing, what with attempted changes in laws, real changes and legal wrangles over interpretation of existing laws. The European Patent Office in Munich took advantage of the lack of agreement over interpretations of existing rules in granting what appears to be the first European patent on plants before any new regulations had been accepted. The patent went to a US-based biotechnology company, Agri-genetics of Boulder, Colorado. The patent protects both a technique for increasing protein content of forage plants such as alfalfa, and the plants produced. By 1988 the Munich Office was said to have hundreds of applications for plant and animal patents, and no new laws had yet been made.

As Pat Mooney reflected, *in the history of the patenting system, people keep losing out.*

Plant Breeders' Rights was the first great loss, Pat Mooney concluded. PBR allowed the European Commission, industrialists and the European Patent Office to argue that if a degree of patenting of plants is allowed through PBR, why not animals too? Industry argues that since there are precedents to the patenting of life, it should therefore be allowed for any living thing. If plant breeders have had a right to patent protection on the fruits of their labours, then the new kind of breeders should have that right, too.

Using the rationale which gave rise to 'breeders' rights' as the ethical justification for the patenting of living things may sound like neat and tidy logic, but PBR itself is controversial and, once inspected closely, a weak example of patentability. Plant Breeders' Rights have been disputed since their inception. It has been near impossible to establish a legal identity for a plant or variety. Also, breeders are not required, as are other inventors, to come up with an 'inventive step' to secure PBR. (They can't, Mooney points out, because the nature of the science makes such a step impossible.) Hence, in practice the patenting system gives way to allowing patenting of plant 'discoveries', even though this is contrary to the idea that patents are rewarded for inventions involving a creative step, and that discoveries are not patentable because 'the inventor (in theory) cannot show how he/she accomplished the work – no intellectual "trail" is left for others to follow.' It has been the political influence of companies, Mooney states, which allowed the patenting of plant discoveries in the United States and elsewhere. He adds, 'How certain is the US Department of Agriculture that patented varieties are not ripped-off from other parts of the world?' Furthermore, contrary to received wisdom, PBR has not stimulated progress in breeding. In fact, 'PBR acts like a noose around the neck of scientific progress. The pressure to breed that which can be patented is such that the development of new breeding techniques and approaches cannot be pursued if they do not lead to patents. In effect, *rigor mortis* has settled over plant breeding.'[9]

There were overt challenges to the principle of plant breeders' rights as a basic human right. Mooney relates that a Canadian Royal Commission came out against PBR in 1960, and a few years later a US Commission recommended abolishing the Plant Patent Act of 1930. In Europe, PBR was successfully attacked in the courts. The consequences of PBR and the ensuing patenting of plants and other living things are far-reaching and do more than protect the interests of the inventor (here, breeder). The protection of the inventor's 'rights' here also acts against the interests of the community at large. However, corporate interests have prevailed. 'Only when farmers and consumers are aroused, do politicians become concerned,' Mooney found.[10]

Such was the situation in the 1980s, as coalitions of farmers and public interest groups forestalled the change in law that would allow patenting of living things in Europe, although how long this would last was not certain; the industrial lobby was fighting for acceptance of the European Directive proposals on patenting living things.

Extending the Boundaries

As well as plants, micro-organisms or rather 'microbio-logical inventions' have long been the subject of debates over the boundaries of what is and what is not patentable in the living world.

Louis Pasteur was granted a patent in microbiology in 1873 by the United States . The patent included a claim to a biologically pure yeast culture, but it was one of only a few patents allowed on micro-organisms. Generally, a 'product of nature' principle prevailed. Because micro-organisms are found in nature, they cannot be inventions. Acceptance of the patentability of micro-organisms was probably slowed down by their ambiguous nature. A strain of bacteria, say, is not fixed in structure or characteristics; it is a living system, capable of changing. Accordingly, inventors had to be satisfied with protection of a *process*, not the living substance, at least until recently when changes in patent laws in many countries 'seem to leave the possibility open'.[11]

In 1969 the Supreme Court in the Federal Republic of Germany moved the boundaries, deciding that the methodological use of biological forces is such a techno-logical process that patenting could be considered (though existing regulations prohibited it). Call it technology and it becomes patentable.

Not the first, but the most influential overturning of the 'product of nature' principle occurred in 1980 with the decision of the US Supreme Court in Diamond v. Chakrabarty, granting Ananda Chakrabarty, a scientist at the General Electric company, a patent on a strain of bacteria which was genetically modified to feed off oil slicks. In a five-four decision, the Court overturned the ruling that a bacterium could not be patented. They argued that the patentee had produced a new bacterium with markedly different characteristics from any found in nature, and that it had a significant utility. The decision read, 'His discovery is not nature's handiwork, but his own.'[12]

With a turn of phrase, a twist in the logic, micro-organisms are now patentable. The decision spurred investment in commercial genetic engineering research in the United States, and it was directly linked to the start of at least one biotechnology company, Genentech. The decision was rapidly followed by further changes in the definition of invention and patentability.

- 1980–84: basic genetic engineering processes were patented by Stanford University and the University of California San Francisco as a result of the work of Stanley Cohen and Herbert Boyer. The US universities gained genetic engineering grants worth an estimated US$1 billion.

- 1984: at the University of California Los Angeles, a patent was granted on a human cell line. (A cell line is a cell culture derived from a single progenitor cell.) The original cells from which the cell line was derived were removed from the spleen of one John Moore, a patient being treated for a form of leukaemia. The value of the cell line was estimated to be worth US$3 billion. The chief researcher involved and the doctor who took out Moore's spleen, David Golde, eventually signed up with two companies, Genetics Institute Inc. in Massachusetts and Sandoz Pharmaceuticals of New Jersey, to develop the cell line into a commercial product for cancer treatment. John Moore says he was not informed at any point, but when he found out, he sued, arguing that the cells were his, not the property of the researchers. The case was still in court in 1990. A 1986 survey showed that in almost every case of tissue usage, patients have not been informed that they have contributed to research. The basic question the suit has raised is: Do we own our body parts? Do we have property rights to our own bodies?

- 1986: the first human vaccine made by recombinant

DNA technology, a hepatitis B vaccine, was approved for marketing. Patent disputes (which are common these days among biotechnologists) erupted. Although patents governing the *process* in making the vaccine were held by the University of California, by Chiron Corporation and by the University of Washington in the US, French researchers at the Pasteur Institute in Paris disputed the marketing. Hepatitis is caused by a virus, and hence its vaccine is made up of a virus or parts of a virus. The Pasteur Institute researchers had a claim to patent rights on the viral DNA, the genetic material – a recent and weird category for patent protection.

- April 1987: the US Patent and Trademark Office passed a decision to allow the patenting of genetically engineered non-human multicellular living organisms, including animals, with the granting of the first patent on a modified animal, an oyster. The oyster had been genetically engineered to have multiple copies of the normal number of chromosomes, making it sterile, and allowing it to grow bigger and tastier. (The argument goes that instead of having to expend energy on reproduction, the oyster spends its energy on whatever it is that makes it taste good.) Although human beings are not patentable in the same way, analysts say that the law could be interpreted to cover human traits.

- February 1988: the US Patent Office declared that farmers who breed genetically altered animals covered under US patents (such as the patent-protected oyster) would have to pay royalty fees.

- 13 April 1988: legal history was made when the US Patent Office issued the first patent on a mammal, a genetically modified mouse engineered to develop breast cancer. The cancer-prone mouse was created by Dr Philip Leder and Dr Timpthy Stewart, scientists at Harvard University. They succeeded in inserting a modified human gene into mouse eggs that would make the resulting mice susceptible to cancer. The mouse patent covers a whole strain of mice: the genetically engineered mouse and all its progeny and all their progeny for 17 years. Harvard University granted an exclusive licence to market the mouse 'invention' to the chemical company Du Pont of Wilmington, Delaware, which had given the researchers US$6 million in research grants for first right of refusal. These mice were the first genetically engineered transgenic animals to be sold commercially. At the time the patent was granted, there were between 20 and 30 other applications for transgenic animals pending at the Patent Office. Biologist Ruth Hubbard and DNA historian Sheldon Krimsky asked, 'Can Drs Leder and Stewart really be said to have manufactured generations of mice because they operated on the eggs of one? . . . The genetically engineered mouse is reputed to be an important contribution to research in understanding and treating breast cancer. This would truly be an exciting outcome. But the patenting of the mouse is no more related to the advance of scientific research than advertising is to the quality of a consumer product . . . university patenting is a symbol of a new era of proprietary knowledge in the health sciences. Harvard University not only holds the prize of the first patented mammal, but it also leads the nation in the number of biomedical scientists with ties to the biotechnology industry: at least 60 Harvard scientists are associated with 33 different firms.'[13]

The patenting system pinpoints the remarkable extent to which biotechnology has consolidated the partnership between academic science and industrial science. It also illustrates the changes that are taking place in the organisation of biological research. Academic researchers have been the major source of DNA-related 'inventions' and cell line modifications used in research and diagnostics. Harriet

Zuckerman, a sociologist at Columbia University, comments, 'The importance of all this is plain. The communication of scientific contributions by academic scientists is apparently becoming less open than it was.'[14] Increased secrecy is a result of the patenting system: if anyone else uses the idea before it is patented, it is no longer patentable. And, as many commentators point out, secrecy is not in the public's interest. Academic researchers – many of whom would insist that they believe in the ideal of a free flow of scientific discoveries for the sake of scientific progress – are none the less increasingly becoming involved in the secrecy game that surrounds patenting. The transnational Ciba-Geigy has even tried to turn the argument around, advocating the position that protection of 'intellectual property' serves the public interest because without it industry would not invest in public research.

In the history of the patenting system, people keep losing out.

Name that Organism[15]

In Europe, the road to patenting life has been more difficult than in the USA or Japan where patenting of living things except human beings is now allowed. The European Commission is trying to follow suit and expand the terrain of what is patentable to suit 'biotechnological inventions'. Just as in their treatment of plant patents, they tried to force the issue through changing the language. They simply changed the definitions of things. Terms such as organisms, parts of organisms, products of organisms and processes were expanded or contracted to suit a criterion of patentability.

For example, the European Directive proposing changes in the patenting law said: Any biological *processes* carried out with the use of *micro-organisms* and involving human intervention should be patentable. The trick was to define the terms *processes* and *micro-organisms*. Micro-organisms have

traditionally meant bacteria, viruses, algae, fungi, rickettsiae and protozoans. The European Directive suggested
that micro-organisms were all of these things and more:
that *all* cells, and apparently human cells and DNA, were also
micro-organisms. The addition of any sort of cells and
genes gives the term a totally new meaning. The Directive
additionally stated that parts of organisms are also patentable, such as cells, genes, and plasmids (circular pieces of
DNA that are not bound up in chromosomes). On the
human front, this would presumably leave egg cells, sperm
cells, and embryos (groups of cells) open to patenting,
although it would not be surprising if somewhere along the
line these will be considered special cases warranting
different rules.

As the term micro-organism was expanded, that of
biological was contracted. In patent law, processes are
techniques to obtain a certain product. The European
Patent Convention of 1973 explicitly excludes the patenting
of 'essentially biological processes' with the exception of
'microbiological' processes (those using bacteria or other
micro-organisms). One reason for forbidding the patenting
of biological processes is that they are 'so closely linked to
nature that you cannot call them inventions . . . They can be
used to improve plant or animal production, but not
owned'.[16] The European Directive got around this barrier
by making the narrowest possible interpretation of 'biological': it suggests that every process that has at least one
non-biological step in it is patentable.

Characteristics of organisms are being considered patentable too. In the US the company Sungene patented a
sunflower with a high oil content. 'What has been patented
is "high oil content",' noted Henk Hobbelink at a meeting in
1988 of the campaigning organisation, the Seeds Action
Network-Europe. 'Sungene has written to all its competitors saying that it will consider an infringement of their
patent any sunflower variety with a high oil content.'[17]

Exerting its expansionist claims, the European Directive

on patenting life forms accordingly pronounced in Article 17 that anyone who infringes on the rights of the patentee, even if unwittingly, is guilty. It sounds as if patent protection is getting greedily out of hand. 'It is the combination of product patents, process patents and this extension of process patents to the final product which amounts to foolproof patent protection. Companies will try to put as many different patents on a plant as possible,' protested the Seeds Campaign of the International Coalition for Development Action, a network of over 700 development-oriented groups and agencies in 21 industrialised countries.[18]

Intellectual Property

Patents are just one of the new claims to bio-scientific 'property', its profits and its control. Another is *copyright*. Can copyright laws be used to protect the commercial value of the genetic engineers' endeavours? In 1982, the legal journal, *George Washington Law Review*, presented detailed arguments as to why genetically engineered microorganisms and cell lines might be copyrighted in the United States.

Supporters of the copyrighting idea said that manipulation of 'genetic information' within cells can be deemed works of authorship and researchers should enjoy the same property rights as writers or computer programmers (an act of 1976 allowed copyright protection of computer programmes). Notice the convenient change in vocabulary: they are talking about the manipulation of genes, which becomes 'genetic information' for the sake of the claim. The analogy was taken as far as it would go, even to suggest that the genetic 'information' contained in nucleotides (components of DNA) is transferred into something useful (proteins). At one point the genetic 'work' was compared to a computer programme.[19]

The idea resurfaced a few years later in the context of the controversial 'human genome project', the plans to map and sequence the entire set of human genes. In the United States, Walter Gilbert, the scientist who lost the race to the human insulin gene, founded a private enterprise, Biogen, to undertake the work. He was interested in doing it quickly, without the bureaucratic delays of publicly funded science. 'Doing it privately,' he insisted, 'does require that the information which is created be protected in some way. I generally thought – and still think – that the information in the sequence would probably be protected under the law of copyright.'[20] It is like saying the person who found the Dead Sea Scrolls should have the copyright on them.

'Intellectual property' is about the ownership of ideas and discoveries, as well as that of inventions. It takes in things like copyright and patents as well as some vaguer notions about owning ideas, about recognition and reward for individual ideas and discoveries. It certainly has a long history in science, and reward has been in the form of kudos as well as money. Scientific institutes like the Royal Society (Britain's equivalent to a National Academy of Science), founded in 1660, took measures to offer its members protection and recognition for their work. Scientists of the seventeenth century (then called natural philosophers) deposited sealed and dated accounts of their discoveries with the Society. Today, for example, proper allocation of credit is signalled by the ordering of authors' names on scientific papers.[21]

However, the emphasis behind the idea of granting 'intellectual property' in plant breeding and similar biological interventions is skewed. Pat Mooney notes that farmers and gardeners are just as inventive as industrialists. The concept only works one-way: in the interests of the researcher, scientist, engineer, inventor. The laws of the industrialised world allow genetic raw material anywhere in the world to become the property of private interests, with no recognition of a previous existence for the 'raw materials'. So too with Walter Gilbert's idea to copyright human genes: the human body becomes the raw material in his pursuit of intellectual property.

The latest developments in biological science to formalise recognition and reward are a re-run of some old conflicts over 'intellectual property' (here, the ideas, discoveries and inventions of researchers, scientists, doctors, industrialists) and epitomise the new conflicts brought about by the industrialisation of life.

What Am I?

Proponents of patenting all living things argue that there are legal precedents, it has already been done. They ignore the absurdity of squeezing already existing life into the general patenting criteria. They insist that the reason plants and animals were removed from patent protection in the past is either because they seemed to be protected by some other means (e.g. plants by the PBR system) or because of lack of forethought (as with animals). This is the view of the industrial world. And many lawyers are defending it.

One person arguing in favour of patenting changes in Europe was Sandra Keegan, an administrator within the European Commission who participated in the elaboration of the European Directive on patenting biotechnological inventions. She opted that patenting was a straightforward idea if one makes a simple distinction between a 'concept' and 'the body', between 'material' and 'spirit'. 'Just because material is of human origin does not necessarily mean that it has a particular strong moral significance,' she argued. 'There is little or no association of the sanctity of human life with human parts as such.'[22]

Keegan's argument, with its distinction between 'spirit' (the non-material) and 'material', is based on a dualism in Western thought dating back to ancient Greece. Each century since brought philosophers and thinkers to bear on the problem of how mind and matter are different, if they are at all. The distinction may seem a 'natural' one or intuitive or just common sense, but it is not the only way to perceive our experiences, ourselves and other living things. The abundant theories of the mind/body distinction may be helpful at times but certainly have not always been benign. As a world view, the mind/body distinction takes some of the blame for the global problems we face today. Calestous Juma, in *The Gene Hunters*, recalls that the division between body and soul which buttressed the scientific Enlightenment and European Christianity was the inducement and precept which allowed gross exploitation of the natural world as 'material' devoid of any other worth. He contrasts this belief with that of paganism, which saw the natural world as enspirited. Trees, rocks, streams, everything had a spirit; a belief which, he argues, protected nature from harsh treatment.

Sandra Keegan's view is closely aligned with that of another lawyer, Lori Andrews in the US, who has argued for a body-as-property line in the area of newer reproductive technologies (*in vitro* fertilisation or IVF, removing eggs from women's ovaries, creating human embryos in 'test-tubes', experimenting with those embryos, surrogacy, using foetal tissues from abortions for medical therapies, etc.). She believes that we all should be allowed to sell or rent our bodies and body parts to medical science and any other buyers and sellers in the reproductive market. In fact, she says acceptance of this view is necessary for women's emancipation, a view contested by other feminists, myself included, who do not see the status of women being improved in this way. To paraphrase Susan George: If nothing is done to overcome inequalities and oppressions, the 'right' to buy and sell our body parts is going to make matters worse, not better.[23]

What does such a right to buy and sell our bodies mean? Healthy but impoverished Turkish 'donors' sold their kidneys for cash to the private Humana Wellington Hospital in London, where the organs would be transplanted into wealthy patients. When the incidents came to light in 1989, it was condemned as exploitation of poor people, and legislation banning commercial trafficking in live human organs followed, but dilemmas remain. As organ donation is lauded as an act of altruism, emotional pressure may be exerted on possible donors close to a patient. Also, indirect payments of money and other incentives may operate.[24]

The explicit implications in a mind/body split were brought home to me while talking to an IVF doctor in London, Dr Jack Glatt. Dr Glatt implied that sperm donation by men is equivalent ethically and morally to egg donation by women. How can that be, I asked, considering the administration of drugs and surgery the woman must go through, the whole technology of reproduction behind egg donation. He said, 'It's the concept that counts. The rest is incidental.' Must we conclude that whatever is done in the clinic is inconsequential because it is not women's physical beings which matter, just the concept behind what is being done?[25]

The body-as-property line affects everyone, as it has been extending to all subjects of medical intervention. 'The human body is not what it used to be,' announced a book called *The Body as Property* by Russell Scott (Allen Lane, London, 1981), which described numerous social, moral and legal questions prompted by the use of human parts in transplantation, and in other areas of medical and scientific research.

This brings us back to John Moore's argument with Dr Golde. Moore is rightly aggrieved by the fact that the medical specialist and scientist used his spleen cells without his knowledge, and created a lucrative product from it, to boot. I have sympathy with the instinctive response that his body is his, and my body is mine, and nobody should be allowed to use pieces of them without consent, at the very least. However, I do not think this concept of the body as property is positive enough; it gives us the right to be potential consumables. It leaves the most vulnerable wide open for the same exploitation that Moore experienced, this time with their permission. Exploitation of the poor and dispossessed already happens; some have been forced by their circumstances or cajoled to sell their organs to doctors who use them for transplants. To say we own our bodies does nothing to challenge the philosophy that sees us as repositories of spare parts for use in medical technology. Whether or not you accept the path along which medical science is going, the question of how we relate to ourselves is crucial. Personally, I agree with those who are challenging the supposed need and growing demand for living bits and bodies. I want to know what other medical approaches are being ignored by the exclusive focus on frontier science in the clinics and operating theatres.

I am sympathetic to the concern about how the language used to describe living things is being altered to fit them into the logic of patenting. The language and the categories used in the patenting of so-called biotechnological inventions turn living things into reserves of genes, to be extracted and used as private property. It imitates the language used to describe and utilise inanimate nature. Physics, for instance, has its atoms, its tiny 'building blocks' of matter. Biological engineering applies this view of the universe to living things, seeing genes, cells, tissue and organs as the 'building blocks' of living 'matter'. It is a view which fits perfectly the proprietary claims of entrepreneurs. Marie-Angèle Hermitte, a French researcher in the law on life forms, commented, 'We see a rise of gene merchants, gene salesmen, in the shadow of those who marketed chemical molecules for pharmaceuticals.'[26]

How we conceive of ourselves and other living things is affected by changes in science and technology; the gene revolution might be understood as a genetic ideology in the making. Genes are literally the negotiable entity in the case of patent and copyright protection. The reorganisation of living nature into the raw materials, exploitable resources, the stuff of inventions, touches on something vitally human: how we think of ourselves and our own bodies, as well as the bodies of other living things. Patenting living things and parts of living things arguably increases the arbitrary manipulation of life, including our own.

6   Vaccines and Future Ills

'There is no enemy to be eradicated.'
Frédérique Apffel Marglin[1]

The first approval for the sale of a human vaccine made by gene splicing techniques, in 1986 in the USA, was for hepatitis B vaccine, manufactured by Merck, Sharp and Dohme in collaboration with Chiron Corporation. At the time, it was considered to be worth $100 million a year for Chiron, and to have 'a potential world market of hundreds of millions of doses . . . But at $100 for the course of three injections, the vaccine costs about the same as conventional plasma-derived vaccines and is far too expensive for mass immunization programmes in countries where the disease is most common.'[2] Patent rights to the virus's genetic material were in dispute.

When I first began thinking about biological engineering as a *technology* – as a system and process around which industry, academic research and medicine is being organised, and as distinct from a technique for manipulating genes and cells – there were three attractive prospects which seemed to be made possible only by genetic engineering. They were the ability to create a better kind of insulin for people with insulin-dependent diabetes through genetically engineered bacteria; altering bacteria for use in cleaning the coastal environment of the effects of the obscenely huge oil spills that have occurred over the past many years; and creating a vaccine for AIDS (acquired immune deficiency syndrome). Could not these developments be wholly beneficial, given proper motivation and careful implementation?

There are no easy exceptions that can be considered outside the whole context of biotechnology. Genetically engineered 'human insulin' or the release of novel 'oil-eating' 'superbecteria' may at first appear to be inspired solutions to specific medical and environmental problems. But, as we have seen in previous chapters, they are not the only nor the necessarily appropriate answers to questions such as, 'How can medicine help people with diabetes?', or 'How can the coastal environment be protected?'

But what about possible vaccines for AIDS or hepatitis B? Trying to answer that question, I became aware of controversies over the effectiveness, safety and use of vaccination *per se*, and of disillusionment among those who find themselves up against powerful medical or government authorities when they resist being vaccinated or having their children vaccinated. Vaccine use brings into play scientific theories, economic profitability, global development and, sometimes, military interests.

Genetic engineering and cell engineering have opened another door for vaccine research. More vaccines are being developed and for an increasing number of conditions. The scope of vaccine use is expanding. Biological engineering enhances the capacity for manipulating cells, DNA and micro-organisms necessary for the development of vaccines. In principle, the possibilities for vaccination and immunisation seem endless; and in practice, a great deal of research is focused on creating vaccinations for infectious disease, cancers, tooth decay and even pregnancy (which is not a disease). The use of vaccination is no longer confined to infectious diseases such as diphtheria or polio. Vaccines are now considered useful for dealing with any conditions which are thought to be controllable biochemically.

The history of vaccination, the thinking behind vaccination, its implementation and expanding medical role may at first seem a side issue in a book about genetic engineering. However, it is a good place to begin our exploration of the 'clean revolution' of medical biotechnology. The fact that medical aspects of biotechnology (including genetic engineering) can be called 'clean' – pure, faultless, unsullied by the problems that accompany agricultural biotechnology – makes the exploration all the more important. This chapter, which is about vaccines and immunisation, lays the groundwork for the next, which is about human genetics and 'gene theories' of medicine.

Making History

Several types of vaccines are in use today. Most expose the body to substances associated with a disease – bacteria, viruses, bacterial toxins or parts of viruses – in order to protect the person from diseases caused by those bacteria and viruses. The principle is explained in terms of knowledge of the body's immune system. When the body is exposed to a foreign substance, say bacteria, it may produce antibodies (a type of protein) specifically tailored to the bacteria. The antibodies 'fight off' the infection, and continue to provide long-term protection. Vaccines, which are composed of weakened forms of the bacteria or virus in question, should stimulate the production of antibodies while not endangering the subject vaccinated. Thus a vaccine provides long-term protection or 'active immunity'. A second type of vaccine, such as that used for hepatitis B, dispenses short-term or 'passive immunity'; here the active ingredient in the vaccine is a substance similar to an antibody.

Conventional medical opinion holds that vaccination is effective in preventing disease in individuals and epidemics in the community. If groups of children who have been immunised and groups that have not been immunised are compared, it is shown that those vaccinated have less incidence of the infectious disease. The health risks of vaccines are said to be minor, with very few cases of major health damage to a person immunised. Medical judgment says that the benefits of vaccination to society at large far outweigh the risks. More children would die or be disabled if vaccinations were not given in comparison with the number of chidren who die or are disabled from vaccines.

The theory of immunisation can be traced back to the ancient Chinese and Greeks, among others, who observed that a case of smallpox could protect a survivor from the disease in future. The observations in other cultures and earlier times led to a number of different methods for treating smallpox, among them a method called inoculation or variolation. Inoculation introduced a small amount of human smallpox matter from an infected person into a susceptible person. A widely used method was to prick the skin and expose the cut. There is evidence of inoculation being used in many different places in the world, and it seems to have been introduced into some areas of Europe in the seventeenth century where it was used by peasants. In the early eighteenth century, the practice was brought from Turkey to England by Lady Mary Wortley Montagu, wife of the British ambassador.

At that time in England, there was knowledge among dairy and farm workers that if they became infected with cowpox, which is not fatal in humans, they would be safe from smallpox, which was usually fatal. The English physician Edward Jenner (1749–1823) learned of the farm workers' knowledge from a young woman patient. In 1796 Jenner tested the theory on an eight-year-old boy, vaccinating him with fluid from cowpox pustules which had infected the hands of a dairy worker, Sarah Nelmes. He later injected the boy with smallpox material from the blisters of someone with the disease.[3] That and subsequent experiments were written up in a scientific paper published in 1798 stating that inoculation with live cowpox virus would protect the subject against the dreaded smallpox. Jenner is credited with putting together the two ideas, inoculation with smallpox and protection afforded by exposure to cowpox. He then put into practice his deduction that inoculation with cowpox would be less harmful than that with smallpox. Scientific history was made and Jenner became a scientific hero.

Early vaccines, including Jenner's, were not as successful or safe as was hoped, and smallpox vaccine was controversial. None the less, the British government introduced compulsory vaccination in 1852. A terrible epidemic broke out in 1871-2, following 18 years of a mandatory vaccination programme at which time 90 per cent of the population was believed to have been vaccinated, 'backed up by four years of Draconian punishments' for those refusing vaccination. The death rate leapt. The vaccination was at the very least ineffective in preventing the ghastly outbreak of 1871-2.[4]

Organised opposition to compulsory vaccination led the government in 1889 to appoint a Royal Commission to examine the matter. While the majority of Commissioners reapproved compulsory infant vaccination, two dissented, putting forward their view that other, more effective measures of stamping out smallpox were available: by controlling outbreaks and protecting communities from its introduction by sanitary organisation, by prompt notification of outbreaks in order to isolate contacts and disinfect contaminated areas, and by securing healthy living conditions. The dissenters called for a review of the history of smallpox epidemics.

Today, if you look at the statistics showing deaths as a result of smallpox, you find that smallpox was already in gradual decline by the 1780s, before Jenner's vaccination experiments. In the late nineteenth century, sanitary and other measures were introduced for control of the disease.

Vaccination was not solely responsible for the gradual decline in smallpox.

This is true for a number of infectious diseases. If you look at records of mortality since the late nineteenth century, they show a gradual decline in deaths from typhoid, tuberculosis and whooping cough well before the introduction of their vaccines. Furthermore, there has never been a vaccine against scarlet fever, which has also been in decline and certainly is no longer as feared as it once was. This suggests that infectious diseases, and the pathogens associated with them, lose their virulence over time and, most importantly, as a result of the proper public health conditions. The general conclusion is that improvements in sanitation and nutrition, introduced in the later years of the nineteenth century, have done more to improve public health than any other single factor.

I heard this evidence cited by a veterinarian, Dr John Webster of the UK organisation Farm Animal Welfare Council, to counter arguments favouring using genetic engineering to create an increasing number of vaccines. He argued that any assessment of the effectiveness of vaccines must be considered in the context of those other environmental and social factors which promote health. 'At first sight,' Webster explained, 'the potential for genetic engineering to improve health appears almost limitless through improved vaccines, diagnostic aids and the insertion of genes to confer disease resistance. In fact, most of the diseases of farm animals in the developed world that are amenable to these approaches are already reasonably under control. The diseases that remain are either non-infectious consequences of existing husbandry systems . . . or infectious diseases such as pneumonia and mastitis which are multifactorial in origin and where the vaccinal approach has been singularly unsuccessful.'[5] He emphasised his belief that genetic engineering is neither a 'cornucopia' nor a 'Pandora's box'. He thought each case for using genetic engineering should be judged on its own merits. This is surely not an unreasonable suggestion. Why, then, do medical establishments continue to embrace vaccines as wholly beneficial? On what scientific basis does it rest?

In the early days of vaccination, Jenner in Britain, the French chemist Louis Pasteur and the German bacteriologist Robert Koch (1843-1910), among others, were united in their interest in disease-causing micro-organisms and in creating vaccines against these infectious agents. The invention of the microscope in the seventeenth century made bacteria visible, and thus made their interest possible. The 'germ theory' of disease (bacteria and viruses identified as the germs), which states that different micro-organisms cause different diseases, was well developed by the nineteenth century and gained support and credence from scientists like Pasteur.

The germ theory had its critics then and has its critics now. That may sound stupid. To challenge it may sound like medical heresy. Who can deny that germs (bacteria, viruses, etc.) cause colds, flu, diphtheria? But the germ theory is challenged not because critics do not acknowledge the presence and action of bacteria or viruses, but because it reduces disease to a single 'disease-causing' agent in isolation from the environment in which the germ and the people it affects are engaged, and because it has led to some harmful ways of treating disease.

The Controversy

Childhood immunisation is compulsory in many countries. Medical science holds that any accidents and health risks which accompany it are a small price to pay for the good of vaccination. We are meant to accept that medical assessment, but many people have not.

One person who has not is Walene James, a woman who found herself at odds with the authorities in the state of Virginia in the USA who demanded that her grandson be immunised when she and her daughter, the child's mother, refused immunisation on health grounds. James brought together an impressive body of medical and other professional opinions and evidence challenging the conventional wisdom of vaccination in *Immunization: The Reality Behind the Myth*. Among the arguments which she and others cite are the following:

- Vaccines carry health risks, some of which are chronicled in medical literature. Immediate risks to the person being immunised may be short-term, such as fever and malaise, or long-term and obviously harmful, such as neurological damage and in some cases death. Compensation through the courts has often been awarded to parents for the damage childhood immunisation has caused their children. In Britain, a no-fault system of compensation was set up in the 1970s. It is often difficult to show the causal link between a vaccination and adverse health effect. The medical world is in a more powerful position, considering that they make the diagnoses and they have the scientific back-up which often weighs against the opinion of parents and carers who have lived with a child since its birth and whose informed judgment should be heard, but often is not. The 'good of society' argument is not sufficient to comfort those children and their families who suffer ill effects.

- Long-term effects of vaccination are not fully understood. Some analysts suggest that the increase in childhood asthma, allergy and other chronic diseases in the industrialised world may be attributed to mass vaccinations, among other factors.

- It is not surprising that vaccines carry health risks, considering the substances which go into them. The active ingredient of a vaccine may be whole bacteria or viruses, or parts of viruses which are weakened or killed; or they may comprise the toxins (poisons) the bacteria produce. For example, live virus vaccines include those for oral polio, measles, mumps and rubella, while tetanus and diphtheria vaccines contain bacterial toxins which have been inactivated with formaldehyde. Another type of viral vaccine, called sub-unit vaccine, is comprised of a protein formed by the virus in question. Vaccines for viral diseases are often cultured in human cells or animals, thus presenting risks: non-detectable viruses may find their way into the vaccine. Also present in the vaccine may be residues of substances used in its manufacture, such as aluminium found in types of whooping cough vaccine. These animal and human blood products, or chemical residues from the production process, may cause adverse reactions, including allergic reaction, after inoculation.

- Vaccination is often ineffective in providing protection against the disease in the individual immunised; it does not necessarily give life-long or even any protection..
- Finally, there is the irresistible argument: nobody ever knows if it was the vaccination or a natural immunity that prevented a person from getting a disease.

Those challenging immunisation argue that there are safer alternatives to vaccines. The promotion of greater public health and the raising of living standards are two factors which would allow a child to acquire general immunity without the risks of vaccination. It is vital to achieve greater levels of public health and safety at this time when, for example, public water supplies in industrial countries are now contaminated with industrial pollution and waste.

The issue here for me is the politics of healing. I do not wish to carve up opinion into those who are 'for' and those who are 'against' vaccination; or to try to 'prove' whether or not vaccination is sometimes appropriate or always inappropriate. I want to focus on the disquieting determination of governments and medical and scientific establishments to minimise the risks of vaccination; and I wish to look at the scientific theory behind it, for it is a small step from the germ theory of disease to the gene theory.

In Britain, public distrust of vaccine safety was aroused in 1974 when Dr John Wilson of the Great Ormond Street Hospital for Sick Children publicised a study of children with brain damage, showing a strong association with recent vaccination for whooping cough. The drug company Wellcome, major supplier of whooping cough vaccine in Britain, has since made an effort to undermine such an association. When the type of vaccine manufactured by Wellcome was associated with brain damage, Dr Arlwyn Griffith, a chief scientist in the company, softened the correlation by coming up with the 'trigger theory'. He argued that Wellcome's 'whole cell' vaccine (containing whole bacteria) did not *cause* irreversible brain damage, although it may *trigger* brain damage in a child who would have, in any case, been prone to it. (The trigger argument itself suggests that vaccinations may be of greater risk than their conventional portrayal suggests, but the theory was used to uphold that the problem was not in the vaccine, but with the individual child.)[6]

The Swedes had long since stopped using 'whole cell' whooping cough vaccine. They replaced it with an acellular vaccine made from bacteria-produced toxin. But this apparently safer vaccine may have problems of its own. In 1989 it was reported to be associated with the deaths of 3 out of 1500 Swedish children taking part in a clinical trial. This was not the first time deaths had been associated with this type of vaccine, but the correlation is difficult to prove, as in so many other cases. To confirm a statistical link, there would need to be a much larger trial; the leader of the Swedish research team, Per Askelof, believed that this would be unethical.[7]

More research is supposed to lead to vaccine progress. This case highlights a problem. In the constant search for safer vaccines, someone, often children, often people in developing countries, must be guinea pigs. (In July 1987, an agreement between the governments of India and the USA was signed to enable genetically engineered vaccines for human infectious diseases developed in the USA to be tested on people in India.)

What about AIDS, though? It is very different from other infectious diseases we have known, and much attention has been given to modes of transmission of the virus (sexual behaviour, the sharing of needles among drug users, contaminated blood used for transfusions). These aspects are, of course, important to recognise and act upon. But so are any other possible factors which increase the risk of getting AIDS, the syndrome, from exposure to HIV, the virus. AIDS is not going to go away with improved sanitation. But because AIDS is a disease of the twentieth century, we need to look carefully at conditions of this century to see which, if any, coincide with AIDS. Some commentators suggest that immune difficiency which may lead to AIDS coincides with environmental factors: immuno-suppressive factors may include poverty and lack of sanitation, certain diseases and infections which are known to suppress the immune system, the use of immuno-suppressive drugs such as antibiotics, drug use and overexposure to radiation. We just don't know enough about these. But we do know that many people who have been exposed to the virus and are HIV positive (their blood contains the antibodies to the virus) have not become ill with AIDS. Though some may later do so, others may not.

I am no expert on AIDS. My point in discussing it here is to question the blanket statement that a viral vaccine for the HIV virus is the best possible path to its control. The bottom line is that while such vaccines may prove effective in future, they may prove far less safe and effective than promised, and that in pursuing them so forcefully, we are limiting our understanding of a disease about which we need all the understanding we can possibly come by.

Vaccine Development

Mass vaccination is high on the list of goals of development agencies. UNICEF, the United Nation's Children's Fund, is one of many which promote mass childhood immunisation. UNICEF estimated that the rise in childhood immunisation in developing countries, which they put at 50 per cent by 1989, has saved one and a half million lives and prevented 200,000 new cases of polio. But actual situations begin to suggest that vaccination cannot be disengaged from the context of people's lives, as the following story of the 1988 cholera epidemic in Delhi, the capital of India, illustrates.

Between June and August 1988 the cholera epidemic in Delhi killed over 1500 people. The source of the cholera bacteria was polluted water from shallow pumps. Contaminated water was one factor contributing to the poor living conditions in the slums where six million out of the eight million people in Delhi live. Deterioration of sanitation, lack of medical aid, dust, squalor and malnutrition fuelled the cholera epidemic.

An independent investigation of the epidemic was carried out by eleven representatives from health fields, including Dr Mira Shiva of the All India Drug Action Network and Dr J.P. Jain. They found criminal negligence by local authorities in the sinking of shallow pumps, and an overall lack of attention to the conditions in the 45 government-authorised slum colonies and the 652 unauthorised ones. (Water from a shallow pump is more likely to become contaminated, more so when sanitary conditions are poor; water drawn from further below the ground surface has the advantage of being filtered through many layers of soil and it is further removed from sewage). Documentary evidence showed that the authorities were well aware of the growing rise in cholera cases before the epidemic erupted in full force. The report of the investigating team, *Crime Goes Unpunished*, further showed that the authorities acted irresponsibly upon learning of the epidemic. Instead of setting up local first aid centres and care for victims, the government initiated a useless and dangerous mass vaccination campaign, subjecting four million people to cholera vaccine, at a conservatively estimated cost of Rs40 million. The vaccination campaign was enacted despite published evidence and medical opinion against the practice. Before the 1988 cholera outbreak an official government committee recommended that the practice of cholera vaccination be discontinued; that mortality can safely and effectively be reduced to 1 per cent by rehydration therapy (the replacement of fluids and salts: dehydration is a major cause of death from diarrhoeal diseases such as cholera); and that cholera vaccine was known to present health risks, especially polio and hepatitis. On behalf of the investigating organisation Nagrik Mahamari Janch Samiti, Dr Jain explained that hepatitis B is transmitted by blood to blood contact, and 'in a setting where 4 million people are inoculated with needles which are not properly sterilized, there is bound to be a transmission of this infection.' Regarding the risk of polio, several provocative or risk factors have been found to precipitate an attack of paralytic polio in individuals already infected with polio virus. These factors include, among other things, intramuscular injections and the administration of immunizing agents. As expected, a rise in the incidence of polio was recorded during the mass vaccination campaign.

Dubbing the government's response Operation Mismanagement, *Crime Goes Unpunished* concluded:

the slums and resettlement colonies have become the breeding grounds for diseases, disability and death. The cholera epidemic has only exposed these grim realities ... [and] ... We are convinced that the decision to administer mass cholera vaccination was as much politically motivated as was the sinking of shallow hand pumps.[8]

The government appeared to be responding dramatically

with the implementation of a vaccination programme, yet what was needed were deeper pumps and other sanitation measures, measures for which there already existed written guidelines, but which the authorities ignored. Officials wanted to be seen to be taking drastic action to improve the situation without admitting responsibility for it (their responsibility would have been plain if they had belatedly fulfilled appropriate sanitation requirements in response to the epidemic). When a number of medical professionals expressed disapproval of the mass vaccination scheme, their views were ignored or suppressed.

Crime Goes Unpunished indicated layers of political and ideological complicity, and asked if the vested interests of transnational corporations had put pressure on government. They found the medical system accountable too: 'Bad enough is our medical education system based on the western model which ensures alienation of the medical personnel from the health and social realities of our people and our country.'[9] The report reiterated that the single most important reason for decrease in mortality from infectious disease in the industrialised world is improvements in public health: in sanitation, hygiene and nutrition. The tragedy of the cholera epidemic could have been avoided, they said, if this knowledge had been translated into policy.

The political dimensions of the cholera epidemic are particular to the incident, but they are also a lesson in how political interests can steer the role of vaccination in health care and health development. The selective use of information, the uncertainty (does it work?), the safety – all these questions must be asked in that political context. For example, UNICEF's report 'The State of the World's Children 1989' disclosed that progress towards decreased child mortality is in decline. The reason is that children are still living in poverty. The development goal of decreasing poverty had not been met because of the debt crisis. Indeed, many governments of countries in Asia, Africa and Latin

America have been falling further into debt and are committed to paying back huge sums to the lending banks. Spending on services most needed by people living in poor conditions has been reduced. Malnutrition leaves children vulnerable to diseases like measles. According to this report, poverty, as a result of the slowing down of development projects in the 1980s – projects which would have raised standards of living – is the cause of the deaths of at least half a million children under the age of five.[10]

In a presentation at the international women's conference on health and reproductive and genetic engineering in Comilla, Bangladesh in 1989, Shila Rani Kaur used the term 'technical development exercises' in comparing vaccination and population control campaigns.[11] She contrasted these approaches with the unmet need for sanitary progress and integrated primary health care services. Many governments and mainstream aid agencies have concentrated on seemingly cost-effective interventions such as childhood immunisation and 'family planning'. While family planning is purported to have the benefit of improving women's health, the contraceptives being given to women are often dangerous, more so because of the way they are often administered with a 'top-down' approach which takes little or no account of the women, their lives, their social and economic conditions.

The comparison she makes and the conclusion she draws are very important, I think. While it is true that the health status of women has improved in some areas where 'family planning' policies based on permanent contraception have been introduced, it must be put in the whole context of health care provision, the causes of maternal morbidity and mortality (illness and death, respectively, related to pregnancy and giving birth), and the harm each method of contraception might cause to women. There are two points to address. The first is the idea that the health benefit of decreased pregnancy far outweighs the adverse health effects from the contraceptives. However, the areas of the world to which this argument is applied are those same areas where adverse effects of contraception such as the Pill, intrauterine devices (IUD), long-lasting hormonal injectables and other such methods are magnified for women whose standard of health care is low, and who have few if any medical check-ups before, during and after receiving such contraceptives. In addition, these contraceptives are not the only known 'solution' to the health risks of pregnancy. The Amsterdam-based Women's Global Network on Reproductive Rights listed known causes and consequences of maternal morbidity on which medical and social efforts could be concentrated depending on what possibilities are available. Among the basic causes of illnesses related to pregnancy and giving birth are: anaemia, malnutrition, ignorance, unwanted pregnancy, mutilating practices, unsuitable medical practices such as Caesarean births, lack of services, care and hygiene. These might lead to: prolapse, infection, haemorrhage, eclampsia, obstructed or prolonged labour, which in turn have consequences including pelvic inflammatory disease (infection and inflammation of the pelvic organs which is caused by bacteria or viruses), ectopic pregnancy, infertility, spontaneous abortion, urinary infections and uterine rupture. The basic causes of maternal morbidity and mortality can be alleviated in the Third World, as they have been for women in developed countries (but not for those women living in poverty), through adequate standards of living and appropriate medical care.[12]

The second point is related to the first. Indian researcher Malini Karkal, former head of the Department of Public Health and Mortality Studies from the International Institute for Population Sciences, reminds supporters of family planning programmes that two studies published in the 1970s established that family planning programmes accounted for a small percentage of fertility declines in the 94 developing countries under study. To a far greater extent, improved social and economic factors (raising women's social status) accounted for the decline. She stressed that making contraceptives available and creating the conditions for their safe use is not the same thing as enforced family planning programmes at the expense of 'empowering women and bringing about health and nutrition revolution'.[13]

A similar case can be made regarding vaccination as a 'technical development exercise'. As two critical doctors lamented:

> Even the World Health Organisation has conceded that the best vaccine against common infectious diseases is an adequate diet. Despite this, they made it perfectly clear to us that they still intended to promote mass immunisation campaigns. Do we take this as an admission that we cannot or do not wish to provide an adequate diet? More likely, it would seem, there is no profit in the constituents of an adequate diet for the pharmaceutical companies.[14]

Health and safety should be primary concerns in development. Clean water, general sanitation, better living standards, provide foundations for health. The problems we can see in agricultural development are being repeated in health development. Western medical science is obsessed with control over diseases just as agricultural science strives for monopoly control over the environment. It is this mentality which has led to the dangerous overemphasis on technical development exercises, such as vaccination campaigns, with their objective to 'search out and destroy the offending microbe'.

Military Interest

An understanding of the interests vested in vaccines would be incomplete without a discussion of military interests, because vaccine research is the crux of biological warfare research.

Vaccine research is by definition research on bacteria, viruses and other micro-organisms dangerous to human beings or animals. Biological warfare or so-called defence research focuses on these pathogens either as biological warfare agents or to create vaccines for soldiers and civilians. These pathogens are often the very micro-organisms which medical scientists are trying to eliminate as a health hazard, such as influenza virus, Rocky Mountain fever, rickettsia and cholera bacteria. In 1989, ten years after the world was declared free of smallpox, stocks of smallpox virus continued to be held in two centres by the USA and USSR, the two military superpowers. Destruction of the virus had been delayed so that its genetic structure could be studied, a military decision motivated by the fear that the enemy would use it as a weapon.[15]

Military interest in vaccine research in concert with recombinant DNA technology is intense enough to have made some scientists voice their alarm. In the United States, molecular biologist Jonathan King of the Massachusetts Institute of Technology and science historian Susan Wright of the University of Michigan have been lobbying for a moratorium on military research using gene splicing technology, a subject we will return to in chapter 10.

Fear and Eradication

Science is often thought to exist in isolation from social and political forces. Some scientists would say that the use of vaccines, or the corrupting influence of business and military interests in vaccine research, is of a political nature and thus out of their hands. But there is more to the politics of vaccination: medical and scientific theories about vaccination are part of the political picture, to the extent that they inform and influence the use of vaccination to eradicate disease through mass immunisation.

At first, eradication may seem like an obvious and effective medical goal. Who does not wish that people should be spared from terrible diseases? After all, smallpox was eradicated by a global vaccination campaign as far as can be known. However, the goal of eradication is not a straightforward one.

Where I grew up in the USA children were immunised for certain childhood diseases, but there was no vaccine then for measles, mumps or rubella. I remember learning from my parents that having these illnesses and getting over them was part of growing up. It made you stronger. If you never had such an illness, that did not mean you were healthier than someone who did. The explanation was a way of making sense of experiences. It was a wisdom then firmly rooted in medical explanations of the body's immune system.

For most children in the industrialised world who live in health-promoting environments, measles is not dangerous. Some children do suffer long-term or irreversible health damage from the disease, and a number die every year from it. In Britain, according to the Office of Population Censuses and Surveys, six people (four under fifteen years of age) died of measles during 1987. The risk of death from measles is cited as a reason why measles vaccines are necessary – to save these children's lives. The risk is, however, greater for children who live in poor conditions. Robert Sharpe determined that a high proportion of British deaths from measles occur in unhealthy children with pre-existing illnesses such as leukaemia and immune deficiency conditions where it might be expected they would be less resistant to infection. 'And how many of the remaining deaths in "previously normal" children were linked to poor diet and the resulting lowered resistance?' he asked.[16] Children who are most at risk from a case of measles are also the children most at risk from the measles vaccine, which is usually a 'live virus' vaccine. These are the children living in the poorest conditions, such as the inner cities of industrial countries, who are most vulnerable to illnesses and to adverse effects from powerful medications. A child's social class is a crucial factor in the effectiveness of vaccination campaigns, yet is often neglected in consider-ations of vaccination policies.[17]

The United States is held up as a shining example of the success of mass immunisation in the virtual elimination of measles. However, the *New England Journal of Medicine* reported in 1987 that an outbreak of measles occurred among adolescents in Corpus Christi, Texas in 1985 even though compulsory vaccination requirements had been thoroughly enforced.[18] In 1990 it was reported that during the first half of 1989 there was almost a fourfold increase in the number of cases of measles over the same period in 1988; over half of these cases were in children over the age of five who had been vaccinated.

Instead of reassessing the feasibility of the goal of eradication, the authorities recommended a change in the immunisation schedule for measles, mumps and rubella vaccine, from one dose to two doses. Does a two-dose regimen pose more risks or will a double dose work better?

This immediate reaction of the medical policy makers is understandable from their own point of view; they believe in the worth of vaccination. But its repetitive message is disturbing – 'Don't question vaccination; when there are problems, we'll solve them eventually. The goal of eradi-cation of disease is worth it, isn't it?' Is it? There are several reasons to question that worth.

First, eradication of disease is elusive, even from a conservative estimate. Professor Frank Fenner, who took part in the WHO (World Health Organisation) global smallpox eradication programme, admits that there were, 'a very distinct set of features – biological and socio-political – that led to the success of this programme, a repetition with other major viral diseases is likely to be a much more difficult task.'[19]

Second, the health risks of the vaccine are an issue.

Regarding the smallpox vaccination campaign there is some evidence suggesting that live-virus vaccines, such as those used in the smallpox eradication programme supported by the World Health Organisation, have tumour-causing potential and may activate dormant viral infections.[20]

Third, implementation of an eradication programme has high social costs. For one, coercion of individuals and groups who might resist it. Childhood vaccination is compulsory in most industrialised countries, despite many controversies over its worth and safety. In Britain, where vaccination has not been compulsory since 1948, medical policy is definitely moving in that direction. Considerable pressure is now being put on parents – this usually means mothers – to have their children immunised. A report by the Institute of Child Health recommended making immunisation a requirement of school entry as it is in the USA and other European countries, and recent government changes to the National Health Service have ensured that doctors will be paid more if they achieve a high uptake of childhood immunisations. The goal is eradication in Europe of indigenous polio, measles, neo-natal tetanus, congenital rubella and diphtheria by the year 2000.

In a study of the smallpox elimination campaign in India, anthropologist Frédérique Apffel Marglin concluded that the social and political costs of the eradication programme were too high.[21] The smallpox vaccination campaign was accepted by the authorities of India as absolutely necessary; the vaccinations were administered without consulting and empowering the people they affected most. Indigenous knowledge and practices to deal with smallpox were belittled, and the police were brought in to implement the programme, using violence if necessary. Marglin contends that the programme was doomed to fail the very people it was supposed to be helping. The goal of health in development – defined as saving lives – could have been served by a different attitude toward the role of vaccination. It could very well have been administered without coercion, although to fewer people, with the result that smallpox would have been adequately contained, and the medical goal of healing (albeit not eradication) would have been met without paternalistic and authoritarian measures. Marglin's view is supported, I think, by a recognition of the controversies surrounding mass immunisation, such as the worry of health risks of viral vaccines. A fundamental problem, she emphasised, is that the scientific medical objective leaves no room for mutual adjustments. The military mentality underlying the 'search and destroy' mission of eradication exemplifies the scientific mentality which aims to destroy what is feared and what is not understood. Marglin contrasted this Western medical development model to those which take account of the context of people's lives: where they live, their beliefs and customs, and so on. To the Western observer, she says, this may appear rather anarchic and an inappropriate way to go about treating disease in a scientific manner. On the other hand, it does take account of the many factors that contribute to disease, places each person in a relationship with those factors, and never monopolises authority.

Even many Western observers find fault with the scientific model. The military metaphors used for explaining disease and its cures are often commented upon: 'The war that man is obliged to wage against viruses is arguably one of his greatest scientific challenges . . . the war against the virus has many battlefields and many fronts.'[22] The waging of war needs generals and soldiers and weapons. Writing from the US in a similar vein, Walene James commented on vaccination in the context of the whole philosophy of the medical system:

they escalate cost, dependency, and concentration of power outside of oneself. Perhaps, even more important, they promote a sense of alienation from one's own body and the natural world that sustains it. The body is not seen as an extension of one's consciousness and part of a larger continuum that includes all life, but as an isolated object that must fight for survival and defend itself against a hostile environment.[23]

The rhetoric of eradication thus plays on fear; similarly, the eradication of disease is a continuous theme of genetic screening: medical science will eliminate for all time certain genetic disorders. But what if people do not want the vaccines and screenings which are necessary to eradication programmes? Will genetic screening one day seem so necessary to society, so necessary to the potential 'eradication' of diseases, that it becomes either compulsory by law or by social pressure?[24]

This is the place where individual freedoms may be lost, ironically because the rhetoric of screening so strongly focuses on the individual as the 'site' of a medical problem, forgetting what is already known about the limitations of medical theories and the need to recognise all the factors, social and economic, involved in health and illness.

Genetically Engineered Vaccines

Given the existing problems with vaccines, what can be said about the new generation of genetically engineered vaccines? So far most of these are vaccines against viruses. Researchers say they are engineered to be safer and to work better. For instance, instead of using the live virus as the active ingredient of a viral vaccine, genetic engineering will allow researchers to manipulate the virus so that only a part of it will be needed to stimulate protection against the disease in question.

This type of vaccine is called 'synthetic vaccine', made up of a small protein removed from the virus's outer coating. The vaccine is therefore no longer a virus, and the dangers from using whole live viruses will be eliminated. How effective is this? The protein would have to be shown to promote an immune response to the disease; and it would have to be shown not to be harmful. Much experimentation is necessary, testing on animals, and then clinical trials on people, more manipulation of viruses, more focusing of time, resources and energy on specific biological agents, since there is insufficient scientific understanding of the kind of immune reponses each synthetic vaccine triggers.

The first approved genetically engineered vaccine for use on humans, mentioned at the beginning of this chapter, is a synthetic vaccine. But Wheale and McNally stress that synthetic vaccines which avoid the use of harmful viruses in preparation and final product are not now the most sought-after type, because it is more difficult to get a good immune response from these types of vaccine. Hence much more attention is paid to live viral vaccines. All of the contemplated vaccines for AIDS to date are composed of live virus. In short, it is likely that harmful viruses will be used in the preparation and possibly in the final product of many new genetically engineered vaccines against viruses.

The use of living viruses as vaccines poses potential hazards to patients, laboratory workers and the general public . . . Live viral vaccines multiply in the subject and produce a continuing stimulus to the immune system and can be dangerous to subjects who are immunodeficient, such as AIDS patients, who can be overcome by infection even by so-called harmless microorganisms.[25]

Researchers describing synthetic vaccines often point out the dangers of live-virus vaccines to contrast favourably with the safety of the synthetic ones. But when other researchers argue in favour of creating live-virus vaccines, the dangers are rarely brought into the discussion.

These complexities become all the more focused when you consider AIDS and the vast research efforts to find a vaccine for the virus associated with it, HIV (human immunodeficiency virus). Genetic engineering and cell tissue culture has led to volumes of *information* (not necessarily wisdom) about HIV. The virus has been broken down into its constituent parts, chemically analysed, and the parts manipulated. A genetic sequence has been worked out. More has probably been discovered about 'the AIDS virus' than any other known virus. But AIDS is still surrounded by myths and prejudice (within the scientific community as well as outside it), and by 1990 it could be said, 'All the high-tech, high-cost AIDS research in the industrialised countries of the North has so far produced very little in respect of the treatment and cure of AIDS.'[26]

AIDS is not one disease, but a syndrome including many symptoms and illnesses; and there is more than one HIV virus associated with AIDS. HIV virus was uncovered by two scientists, one in France, and the other in the United States. Science accepted HIV as the singular, viral cause of AIDS. But the monocausal explanation is becoming more and more controversial. There is evidence suggesting that HIV in and of itself is not the sole or primary cause of AIDS. If this is correct, then the pursuit of a vaccine to immunise against the HIV virus would be an insufficient answer to AIDS, even if a 'perfect' vaccine were possible.[27] Research on AIDS-related viruses also raises a question about the desirability of the vaccine research (and cure).HIV viruses are retro-viruses, a class of viruses which have distinctive properties. Retroviruses integrate their own DNA into the DNA of the human or animal cells which they inhabit. The viral DNA appears to be another gene to the cell. It may just sit there, or it may be activated. Retroviruses have been implicated as a cause of cancer. For these reasons, these viruses may well be seen as potentially dangerous; yet retroviruses are exploited for use in gene manipulation: they are used to carry new genes into animal or human cells.

Further, the origin of HIV and AIDS has been the subject of controversy since the syndrome was identified. There are many theories, but nobody knows where AIDS began. The theory that caught the most attention is that of the African origin of AIDS, although when inspected, there was no solid evidence to back it up. However, top AIDS scientists and researchers have continually embellished the theory with their own speculations. Why? In *Aids and its Metaphors*, Susan Sontag points to 'the racist sterotypes in much of the speculation about the geographical origin of AIDS'.

> Part of the centuries-old conception of Europe as a privileged cultural entity is that it is a place which is colonized by lethal diseases coming from elsewhere. Europe is assumed to be by rights free of diseases. (And Europeans have been astoundingly callous about the far more devastating extent to which they – as invaders, as colonists – have introduced *their* lethal diseases to the exotic, 'primitive' world: think of the ravages of smallpox, influenza, and cholera on the aboriginal populations of the Americas and Australia.) The tenacity of the connection of exotic origin with dreaded disease is one reason why cholera, of which there were four great outbreaks in Europe in the nineteenth century, each with a lower death toll than the preceding one, has continued to be more memorable than smallpox, whose ravages increased as the century went on . . . but which could not be construed as, plague-like, a disease with a non-European origin.[28]

So much effort has gone into trying to prove the African origin of AIDS, while few scientists have taken seriously another theory which suggests that AIDS is a laboratory experiment which went terribly wrong. Did HIV originate in a laboratory? There are two controversial theories which suggest just that. One of them finds the source in vivisection experiments on monkeys; and the other theory points to cell and virus culturing, that is growing viruses and cells under laboratory conditions.[29]

Regine Kollek, a scientist and staff member of the West German parliament's Committee of Inquiry into gene technology, persuasively argued that scientists were capable of accidentally creating an AIDS-type virus in cell work even before genetic engineering came on the scene. She believes this line should be investigated, not to accord blame to a specific person or organisation, but to find out how such viruses emerge and spread for the sake of people with AIDS and to prevent its recurrence.[30]

The possibility of the laboratory origin of AIDS is based on the knowledge that:

- AIDS-like diseases have been found in laboratory monkeys or apes, but not in such animals in the wild
- retroviruses which occur in animals do not 'cross over' to infect the human species; but 'cross-over' becomes possible with experimental methods that break down species barriers
- laboratory researchers had the capability to alter cells and viruses at the time when HIV is thought to have emerged (not by recombinant DNA technology, however, since HIV has been traced back to well before gene splicing was established)

My point in raising these contentions is not to suggest that any one theory is closer to the truth than the others, but rather to suggest that there are uncertainties in the scientific 'facts' about AIDS; that the attitudes and biases of scientists are part of the issue; and that there are many different avenues of enquiry which can be followed. Why some are, and why some are not, is not simply a matter of which appears most scientifically sound. Gerhard Hunsmann, Director of the German Research Centre on Primates in Göttingen, who does not agree with the African origin of AIDS hypothesis, remarked about the situation, 'Naturally it's always easier to sell some explanation than to admit that one doesn't have one. I think this is true of scientists as well as of newspaper editors.'[31] The laboratory origin of AIDS explanation is not very 'sellable':

VACCINES AND FUTURE ILLS   165

after all, what would it say by implication about biological manipulation of cells and their components?

Considering all that has been said about the efficacy, safety, social costs, motivations and limitations of immunisation, how fruitful does the growing list of vaccines appear? The germ theory perhaps had its uses; vaccines perhaps have their uses; but they have both been far overstated and overvalued. Why then the continuing stubborn attachment to vaccination and to germ theory?

From Germ Theory to Gene Theory

As an explanation, the germ theory has been attractive scientifically. The focus on disease-causing microbes and vaccination against them is 'simple', the hallmark of science. However, medical science is at the point where the germ theory can no longer explain everything. Infectious diseases all differ with respect to their epidemiology and their sources. For example, viruses are very different agents from bacteria, and far more complex than can be explained by the germ models of infection. (Viruses are seen as genetic parasites; they take over a cell's DNA. Bacteria do not infect in the same way.)

The germ theory was designed to embrace all infectious disease. It cannot because it is limited. Science has, therefore, revised the concept (without questioning the limitations of the germ theory), and given us a new umbrella theory: the molecular theory of control. 'Molecule' is the term chemists and biochemists give to basic units of matter. Molecular biology aims to study the structure and function of the large molecules of life – primarily DNA and proteins such as insulin, or the antibodies of the immune system – in living cells. This is the approach which has been successful in explaining the role of DNA as the genetic material. Molecular biology ultimately aims to explain as completely as possible in *molecular* terms the biology of organisms. It is not so much concerned with how organism and environment interact, or with the body as an integrated system, although ultimately it must recognise these aspects of biology. Medicine has, to a large extent, embraced molecular biology and its molecular explanations of disease.

Molecular biology provides the perfect elegant symmetry of causation and cure – in theory. For the past decade it seems every chronic illness has been linked to the individual's faulty immune system (and ultimately faulty genes). Immune system explanations fit molecular medicine perfectly. On the one hand, biological engineering allows researchers to examine all those viruses, bacteria, biochemicals (antibodies) and DNA. On the other hand, the cures – vaccines and drugs – are increasingly the products of genetic engineering. Biological engineering is not just presented as a 'tool' to create more vaccines. It is seen as an approach for the understanding and control of disease. For some wider implications, let us turn to the subject of vaccines against pregnancy.

Anti-pregnancy Vaccines

When control of pregnancy and childbirth moved from the world of women to the profession of medicine, the disease model of pregnancy began. Birth became the province of (male) doctors and hospitals, with (female) midwives playing a less important role, or no role at all, in the care of pregnant women. In this world, pregnancy became a medical condition, not only in those cases when something went wrong, but at all times. (If you look at the emerging medical speciality of sports medicine, you see a very different relationship between doctor and patient. Sports medicine treats people with injuries or deals with other medical aspects of sport playing, be it football, skating or tennis. Professional players and teams often have their own doctor on call. Never does the doctor become a specialist or an authority on the sport itself, an expert on the rules of play. Yet in obstetrics and gynaecology, doctors have terrific authority over the social aspects of female fertility – abortion, for example.)

Women's health movements have shed light on the misogynist philosophies and practices underpinning many of the modern medical scientific approaches to female reproduction, and the relationship between women's individual loss of power, control over fertility and sexuality. There has been a movement towards approaches to childbirth and the principles of midwifery which respect women.

However, the disease model of pregnancy is not dead. It is in the astounding idea of anti-pregnancy vaccines that the analogy with disease is most apparent. The idea of anti-fertility vaccines comes from immunology, where various events in fertility and conception are understood in immunological terms. Immunology stresses the molecular events of fertility, processes between egg and sperm, woman and foetus. For instance, certain causes of infertility are explained immunologically: men may make antibodies to their own sperm; or a woman may make antibodies to the man's sperm.

Anti-fertility vaccines are seen by many population planners as a greatly needed addition to the available contraceptives. They argue that *if* a safe, effective and reversible vaccine could be developed, it would have the advantages of long duration; it could be administered by paramedical personnel; and its components could be manufactured on a large scale at low cost. But, they admit, the novelty of immunocontraception (as it is called) raises new questions regarding the nature and extent of safety studies required at various stages of vaccine development. Potential hazards to the subject vaccinated include toxic effects and, when such a vaccine is administered to women, foetal damage if pregnancy should occur.[32] (In principle, contraceptive vaccines could be administered to men as well as women, but like the male contraceptive Pill it is an idea whose time never really comes.)

Ana dos Reis, a Brazilian doctor, looks at anti-pregnancy vaccines in the context of population control policies. What does inducing biological rejection of pregnancy really mean? she asks. What kind of ethic will authorise its general use? Who is going to determine this ethic? Population planners think in abstractions, in terms of 'populations' where women do not enter the picture as individuals with autonomous lives. In the past, population planners have used statistics to say that using the Pill or other devices is safer than having babies, to 'prove' that contraception 'protects' women against pregnancy. Pregnancy (the population 'problem') becomes an epidemic; hence, vaccines are produced to counter the epidemic. Dos Reis cites medical language which describes contraceptive vaccines as providing an 'immunological attack' on fertility through the 'manipulation of . . . maternal immune recognition system'.[33]

Different types of anti-pregnancy vaccine are being developed in laboratories all over the world. One which is considered promising interferes with a protein produced when sperm and egg combine at fertilisation. When antibodies to the protein are produced, pregnancy is blocked.

A second type is based on a hormone of pregnancy, human chorionic gonadotropin (HCG). HCG is necessary for implantation of the embryo in the woman's womb. An 'immunological attack' on HCG might be expected to interrupt pregnancy at the earliest stage. To provoke an immunological response in the woman's body, the vaccine consists of HCG (actually a sub-unit of HCG called beta-sub-unit) united with a carrier such as tetanus toxoid, the weakened poison from tetanus bacteria. It may be administered with a third component called an adjuvant, a substance which helps provoke the immunological response. The development of each of these components is considered a significant scientific achievement. Immunisation with such a vaccine should neutralise HCG in the woman's body, thus preventing pregnancy, although the exact mode of action of the vaccine is not clear. The contraceptive should last a certain amount of time (say, a year), but not indefinitely; and as Ana dos Reis reports, 'immunological stimulation is known to provoke permanent responses, even if not detected by normal techniques.' Could this form of contraception cause later infertility for some women?

Indian scientists play a major role in research and development of contraceptive vaccines, which the government there has considered a top priority research area. By one account the original idea of immunocontraception came from Professor G P Talwar, director of the National Institute of Immunology (NII) in Delhi, one of India's showcase scientific research institutes.[34]

By 1989 there were five different anti-HCG vaccines known to be on trial: two from NII, one sponsored by WHO; one sponsored by the Population Council; and one in China. One of these vaccines was being tried out on women in Australia in 1984, and, according to dos Reis, the medical literature makes references to earlier trials on women in Sweden, Finland, Chile and Brazil. In India, NII's first anti-HCG vaccine trial took place in 1986 on 103 women who had previously been sterilised, and who were paid for taking part. The reason for using it first on women who had been sterilised was to ensure that none conceived, since the researchers were not sure of the vaccine's anti-fertility action. Success was judged by analysing blood samples to see if antibodies to the vaccine were present in the women. If antibodies to HCG were found, then the researchers judged that the vaccine worked and that pregnancy could not have developed. For the trial, the women received three injections, one month apart, followed by a booster at six months.

The scientific paper on the first NII trials concluded that the vaccine worked (that is, antibodies were detected in the blood) in most women for up to a year, and in some for six months. It reported some 'minor reactions' at various stages of the trials. These included fever, pains in joints and 'hypersensitivity', attributed to the carrier tetanus toxoid. Dr Charanjit Babra, a WHO consultant on anti-pregnancy vaccines, said that two women from this trial had pituitary complications, but that the researchers did not name these as being related to the vaccine.[35] Why not? Dr Ana dos Reis reports that 'pituitary atrophy' is one of the adverse effects, effects which, she adds, are always minimised in trial reports. It is certainly not unlikely that the pituitary gland, being of central importance to the body's hormonal system, may be affected.

There is a general confidence in the medical establishments that the anti-pregnancy vaccine will prove safe for women, and much safer than existing hormonal contraceptives such as the Pill and long-lasting injectables. According to Dr Babra, anti-fertility vaccines should prove safer than hormonal contraceptives because they do not interfere in ovulation. Interestingly, Babra said that if the harmful effects of hormonal contraceptives which are known today were known 30 years ago, the use of Pills and injectables would never have been allowed. The problem with the hormonal contraceptives, he said, can be located in their interference with ovulation; contraceptive vaccines do not interfere with ovulation (as far as is known), and because they do not, they will not pose the health dangers which accompany hormonals. What do they interfere with?

Such confidence is incomprehensible considering the health risks that might arise from the use of weakened tetanus toxins, the risks of vaccines in general, the lack of knowledge of long-term effects, and the general admission that a novel form of fertility regulation raises new questions about safety and adverse effects. It is incomprehensible considering that animal studies are being carried out to test for possible chromosomal abnormality in the offspring of vaccinated animals, damage to the female animals' ovaries, and damge to male animals' sperm. It is incomprehensible considering that, according to Dr Talwar, physical damage to male genitals has been recorded when an anti-fertility vaccine based on a different fertility hormone (LHRH, leutinising hormone releasing hormone) was under investigation as a male contraceptive.[36]

In this age of genetics, we can also expect to find a strong genetic connection somewhere. For one, foetal damage is cited as a possible effect of anti-pregnancy vaccines. Another 'genetic problem' is the theory that genetically different populations (women) may respond differently to contraceptive vaccines. (Does this appear to be a hyper-technological way of saying every individual is unique?) In Brazil, plans to test this theory were stopped by women's protests. Elsewhere, researchers have suggested that in future, women could be genetically screened to find those who would respond well; or perhaps screening could be used to create designer anti-pregnancy vaccines. Dr Babra believes that genetic screening for this purpose is prohibitively expensive and that it would require an extraordinary amount of research to attempt to make the correlation between a woman's genetic profile and the effect of contraceptive vaccination. It may not seem so outrageous to other scientists, though, considering the often stated belief that by the next century, DNA screening will become so widespread that routine and perhaps mandatory genetic screening will be in place for all individuals, to identify all known markers for genetic disorders and aid medical counselling.[37]

My research on vaccination and immunisation led me down unexpected paths and convergences. I had to ask myself again, where does rhetoric stop and systematic science begin? But that is the wrong question, I think.

This is not to say that science is the enemy to be eradicated. The point is that the political issues of health and health care are not outside science but enmeshed within it. Better to put back the context and admit the limitations of scientific medicine and its power to set up a monolithic theory of disease and medical practice, one that takes little account of the ambiguities of research findings and alternative wisdoms, and one that is often prone to medical bullying and political expediency, not to mention the ties to pharmaceutical and related industries. Better to take a hard look at the ideology of the science.

Perhaps the most important point to make here is that the medicine of prevention is not served by isolating the body and turning it into a molecular battlefield, which just so happens to serve the interests of pharmaceutical and medical supplies companies. Prevention must mean attention to economic and environmental as well as bio-logical factors, and to integrated primary health care. Poverty, nutrition and the debt crisis are as significant as the 'purely' medical questions of disease-causing micro-organisms or damaged immune systems. Medical scientists often see these aspects of prevention as a future goal, not one they can effect. The irony is that each new product of the molecular genetic revolution usually brings more new promises – another future goal, but one that rarely can happen without courting some social or political adversity.

7 Genes-the-Cause

. . . what happens when several genetic factors contributing to susceptibility to cardiovascular disease have been identified, a question now the focus of a great deal of research? It should then be possible to categorize groups of people who are at risk, and who could (and should) take preventative steps, but only by some complicated population screening.

Peter Newmark, *Nature*[1]

Shall we create a special class of genetic outlaws who come slouching through our streets, branded by yellow armbands and with broken double helices advertising their defective status?

Nancy Wexler, *Genetics and the Law II*[2]

One upon a time, the story goes, before there was gene splicing and gene cloning, there were only three ways to study inherited disease, or what was thought might be inherited:

- clinicians and researchers could study inheritance patterns through generations of a family by observing which members were affected
- they could take blood samples to diagnose conditions for which biochemical tests were available, in order to identify persons affected or persons who carry the condition

- they could look under a microscope at cells and chromosomes to look for abnormalities or what they thought were abnormalities

(I take pains to qualify what observers were and are doing. The history of attempts to correlate human characteristics with heredity is riddled with dubious undertakings. For example, an XYY chromosome pattern was linked to delinquency, psychopathology and criminal behaviour in men in the late 1960s and early 1970s. This claim is now generally discounted.)

Genetic engineering changed the scope of all that. Now it is possible to isolate genes (pieces of DNA). It is possible to clone them and study them. Once a gene is associated with a condition, it is possible to use a 'gene probe' to 'search' an individual's DNA to see if they carry the gene in question.

Genetic engineering is the foundation of the so-called 'new genetics in clinical practice'. The 'new genetics' are direct gene analysis and direct gene intervention; the 'old' genetics is based on less direct methods of gene analysis: for example, sickle cell anaemia may be identified by looking at a person's blood cells under a microscope. The 'new' genetics is most interested in studying the fine structure of genes and how they work.

In the previous chapter, I suggested that a gene theory of disease is replacing the old germ theory of infectious disease.[3] This chapter follows through that theory, and puts it in the context of an old idea: that biology is destiny.

Desperately Seeking Genes

There is an incredible amount of genetics research geared towards identifying 'disease-causing' genes and genetic mechanisms associated with diseases. This research is not limited to 'classical' hereditary conditions such as sickle cell anaemia, haemophilia or cystic fibrosis. It includes many other types of illnesses and conditions that are not illnesses.

The expanding scope of genes-the-cause can be seen in this growing number of conditions of the human body and mind which are the subject of gene analysis. During the last few years news headlines have proclaimed:

'Genes implicated in breast cancer'

'Multiple genes for manic depression'

'Pair of genes goes awry in bowel cancer'

'Lung cancer gene located'

'Gene link to schizophrenia'

'Race for cystic fibrosis gene nears end'

'Doctor's locate rogue gene "causing asthma" '

'Single gene may predispose people to allergy'

'Faulty gene leads to old age'[4]

These conditions are each unique and complex in their origins and effects, in their epidemiology – that is, how external influences such as nutrition, infection, toxins in the environment, sex, race, standard of living, and so on may affect their incidence and prevalence. Yet these very different conditions have been gathered under the umbrella of the genetic hypothesis.

Scientific focus on the pathology of genes can be traced to the 1960s with the theory that 'abonormal' genes should be viewed as causal agents of disease, just as bacteria or viruses are considered causal agents; some eminent geneticists proposed that all diseases have a genetic component.

DNA that can be studied in a test-tube or cell is one thing, but how the gleaned information is interpreted is another. When you read scientific commentaries on the actual gene 'finds', it becomes apparent that the discoveries are not straightforward. For example, the gene marker 'found' to be associated with manic depression in a closed community of Amish people of northern Pennsylvania in the United States was not found elsewhere. For another example, the presence in a person of a gene marker associated with schizophrenia does not necessarily mean the person will suffer from or 'get' schizophrenia.

Questions of research priorities and context arise here. Genetic explanations may be the scientific fashion, but as a fashion they diminish those other ways of understanding health and ill health, and appropriate medical care. Although few medical scientists would deny the need to look at the whole context of an illness, including environmental and social factors, the bulk of genetic research does not take these into consideration. Instead the approach is focused at the molecular gene level; then, after a genetic link is 'found', so a cure or therapy will become obvious. The hunt for genetic causes tends to overshadow the substantial evidence regarding other factors in health and illness. If a person's individual biology becomes the major focus of genetic explanations and genetic cures, what are we losing here? Equally important, what do we get from these quests?

The Burden of Genes

Medical students and students of genetics may learn that genetic disease is a 'burden on the community', that the spread of 'morbid genes' burdens the population. Thus any medical intervention which aims to decrease the genetic burden, for instance through genetic counselling, gene screening and selective abortion, is beneficial to the population as a whole, as well as to the individual woman and family involved. This idea of removing 'disease genes' from the population, although not the sole aim of genetic medicine today, remains acceptable among some medical professionals, including some human geneticists.

No such idea is inevitable, so where did this one come from? Genetic screening was introduced after the Second World War, and soon after became known as genetic counselling. One professor of human genetics referred to the 'infiltration of genetics into medicine'. One genetics textbook published in 1977 described genetic counselling as advising patients of the risks of disease to themselves or their offspring if there is a genetic skeleton in the family cupboard'. The authors predicted that medical genetics would one day be used to 'outwit our inheritance' and to 'offset the action of deleterious genes'.[5]

According to historian Daniel J Kevles in *In the Name of Eugenics: Genetics and the Uses of Human Heredity* (Knopf, New York, 1985), genetic screening was an idea which initially came from the eugenics movement, a social movement which was dedicated to the idea of differences in biological 'quality' among individuals. Among its proposed aims was human selective breeding for the 'more suitable' human qualities, through education or compulsory laws or both.

Eugenics always had a close association with the science of genetics. Both emerged at the turn of the twentieth century. Just as the rediscovery of Mendel's pea plant experiments influenced plant breeding and formalised the 'laws' of genetics, so were attitudes towards human breeding influenced. Social reformers who embraced eugenics believed that the empirical findings of genetics could be put to use to breed an innately 'better' human population. Their assessments and ideals speak for themselves. Influential biologist, eugenicist and international figure Sir Julian Huxley wrote in *Heredity and Man*, published in 1937:

> Eugenics seeks to apply the known laws of heredity, so as to prevent the degeneration of the race and improve its inborn qualities . . . Once they have been born, defectives are happier and more useful in these institutions than at large. But it would have been better by far, for them and for the rest of the community, if they had never been born.[6]

The idea did not appear out of the blue, or, for that matter, from new empirical findings. A prevalent idea was that the poor were poor because of some innate (whether biologically inherited or culturally inherited) reason, an explanation which conveniently ignores economic forces and discrimination on the basis of class, race, sex and disability.

Eugenicist ideas were incorporated into repressive laws in many countries. Most well known perhaps are the compulsory sterilisation laws which were enacted in some of the United States in the 1920s, in Sweden, Denmark, Finland and Switzerland, which targeted certain types of people – the poor, blacks, people in mental institutions, namely anyone deemed unfit to reproduce. Eugenics became a dumping ground for discriminatory ideas, and rationalised them as biologically based. It reached a horrifying depth as the basis of the political philosophy of the Nazi 'racial hygiene' or 'genetic health' programme which began as an expulsion or internment of Jews, Poles, gypsies, homosexuals and anyone else deemed inferior, and which ended in mass executions and plans for genocide. To maximise reproduction among those considered of superior and 'Aryan' stock, breeding programmes were set up, with an accompanying ideology of motherhood that elaborated the role of 'Aryan' women as patriotic breeders.

In the aftermath of the Nazi population programme, genetics lost its close attachment to eugenics, but the science of human genetics never completely lost its ties with eugenical principles of one kind or another. Now the bio-revolution gives the word eugenics a new lease of life. The word 'eugenics' is even making a come-back as an acceptable principle.

Proponents of the new eugenics of the new genetics say it is now going to be a good thing based on a good science; this time eugenics will not be oppressive, because it is not about race but about genes which can be observed and measured scientifically; it is not about eliminating certain types of people, but about dealing with genetic medical conditions in any number of ways, for example, creating better diagnostic tests and gene therapies. Indeed the objective of certain forms of gene therapy is not to eliminate 'morbid genes' from the population, but to cure conditions 'caused' by the genes; people with hereditary illness who in the past would not have grown up to reproduce, may in future do so. (The new medical treatment known as gene therapy is discussed further in the appendix.) In fact some analysts would reject the term 'new eugenics', saying genetic intervention today is not eugenicist for these very reasons. All in all, the argument goes, knowledge of genetics and new gene technologies is highly scientific, and should work to improve the quality of life.

However, the genetics of the past also always promoted itself as highly scientific and based on biological facts, but was none the less influenced by eugenicist ideas, and formulated faulty correlations of heredity and behaviour such as the XYY syndrome. And since influential social movements do not disappear as swiftly as eugenics supposedly has, we should judge carefully the present science of medical genetics which proposes that if the presence or absence of a gene is the cause of an ailment, genetic technology may provide the solutions.

Genes and Disabled Lives

The language of medicine tends to hide the person behind an impersonal vocabulary of body parts, while scientific terms tend to mask the fact that when one is talking about a 'deleterious gene', one is saying something about a human being. In the parlance of genetic medicine we hear about 'genetic defect', 'defective gene', 'disease-causing genes', 'chromosomal defects', 'genetic abnormality', 'morbid genes', 'genetic burden', 'genetic risk', 'molecular pathology' and functions which are 'deviant' rather than 'normal', but it still comes down to people's characteristics, and social categories of normality and abnormality. It is easy to lose sight of this in the world of biochemicals, genes and body parts. And it remains that someone must decide which genes and conditions create and constitute a 'burden'. More to the point, someone must decide that genes *are* the problem, the burden. As for the implications: Oxford philosopher Jonathan Glover entitled his book on the subject of the new human genetics, *What Kind of People Should There Be?*

The disability rights movement gave meaning and visibility to a type of oppression with the word 'ableism', discrimination against people with disabilities and the practice which categorises some people as able-bodied and 'normal', and other people as disabled and biologically incapacitated. Like any movement it is many movements, but at the core it questions the definition of disability. Disability is not simply biological; it can best be understood in social terms. For instance the ability to be mobile is not a 'purely' biological condition. The gradation of steps and ramps, the width and height of passageways and tables are all the results of social decisions. By extension, social solutions to problems of access benefit many different groups. A ramp not only enables a person in a wheelchair, it enables a cyclist, a person pushing a pram and someone carrying heavy packages.

Ableism, like other prejudices, is fuelled by misinformation and misinterpretation. Most disability occurs after birth through illness and accident. Less than 5 per cent of disability in adults is genetic in the old 'classical' meaning of the term. Of course if statisticians accept the all-embracing idea that every illness or disability has a 'genetic component', then this statistic would be different, but not particularly enlightening. Who would be so bold as to identify themselves as a perfect specimen? Would short-sightedness be included in such a statistic? The disability rights movement emphasises that any attempt to define biological normality is flawed; there are many ways to be human.

Jill Rakusen, who has written widely on pregnancy intervention technology, finds that in 'The Pursuit of the

Perfect Baby', information from doctors and genetic counsellors is often misinformed, either because doctors themselves know little about disabilities, often because they see only children in hospitals in extreme situations, and sometimes because they fail to mention the uncertainties or complexities of the medical scientific knowledge. She writes:

> To most people, even most women, screening may seem unequivocally an excellent idea. 'It gives women the opportunity to know if the fetus is all right' – is a statement which encapsulates prevalent beliefs. However, despite this ostensible function – which is to reduce anxiety – screening can so easily create further anxieties, and other problems, none of which tend to be given the attention they deserve ... women commonly experience powerful pressures – overt and covert – to have an abortion if screening tests show up a 'positive' result ... little account is taken of their feelings or ambivalences ...

Handicap is much feared in society, she adds, and doctors share this fear with the rest of us; in part, this fear is reflected in attitudes about screening.[7]

In raising this subject, I do not wish to ignore or negate the personal concerns of women who seek prenatal or preconception screening; nor to ignore fears about giving birth to a child with a handicap or illness, but to raise questions about attitudes and accountability in implementing certain kinds of procedures, the whole background as to why, and what it might mean today for people with disabilities and for parents and carers. Activist and lawyer Theresia Degener illuminates:

> fear of the birth of a handicapped child is only too justified because for mothers it usually means isolation and discrimination, giving up their jobs, and/or poverty ... Only too often these economic and social conditions

are confused with the disability itself, by equating it with concepts such as 'misery' and 'despair'.

Another writer on the subject, Anne Finger, disabled by polio, is an abortion counsellor and a reproductive rights and disability rights activist; she uncompromisingly defends the value of 'disabled lives' just as she defends women's lives.[8]

Further, although medical professionals may have genuine concern in giving women an opportunity to control their lives through prenatal or preconception screening, it does not explain why *this* choice of screening is offered, but not other choices such as extended specialised care. One reason is clearly economic, as an argument that is sometimes given, that it is cheaper to screen and select foetuses during pregnancy than to offer the social support and health care needs of children and adults with certain disabilities, shows.

Finally, I am not implying that there is a monolithic eugenicist objective within medical science. The problem is not the existence or otherwise of a secret international conspiracy of eugenicists. The point is that there is a lack of recognition of social prejudices and what can only be called eugenicist ideals and aims within genetic medicine today – that people's lives may beneficially be controlled through genetic interventions; and that these are supporting far-flung theories of genes for conditions as diverse as Huntington's chorea, cystic fibrosis, schizophrenia and susceptibility to occupational illness.

Huntington's Chorea

Huntington's chorea was one of the first illnesses ostensibly to show 'a gene defect link of diseases of the mind'.[9] It is a degenerative disorder of the nervous system; it was known to be inherited long before any genes became associated with it.

People who get Huntington's chorea are usually healthy until the onset of symptoms around the age of 40, when movement becomes uncoordinated and involuntary. They become progressively further mentally and physically debilitated. Huntington's chorea has been described as an 'inherited neuropsychiatric disorder of unknown etiology [origin]'[10]; the 'disease shows itself as a gradual, irreversible deterioration in physical condition and personality.'[11]

James Gusella of Harvard Medical School in Massachusetts led the team which isolated the 'gene marker' for Huntington's chorea by studying a community in Venezuela where there is a high incidence of Huntington's chorea. A gene marker is a section of DNA closely linked to the gene being hunted. Various gene markers have now been isolated for Huntington's chorea by different researchers trying to locate the elusive 'disease gene'.[12]

Like so many quests for the fruits of biotechnology these days, scandal accompanied the race for the Huntington's chorea gene. Gusella refused to give copies of a gene probe he had found to a group of clinical researchers at the University of Oxford who had requested it. The refusal met with some criticism within the scientific community, since sharing of research materials and research findings is in principle considered correct scientific behaviour and necessary for the advancement of knowledge. Gusella argued that he was ethically motivated in withholding the gene marker, on the grounds that any clinical application was premature. On the British end, one of the researchers baulked at Gusella's self-imposed moral and professional judgment, and the implication – that he was the expert on Huntington's chorea and that the British team were too inept to deal with the material.

The incident could have encouraged a useful discussion about the uncertainty of correlations between genes and traits, and between genes and diseases, which was the crux of Gusella's refusal to share his discovery, at least on the surface. Instead, attention concentrated on a scientist's individual rights to own her/his research materials; and its converse, a scientist's obligation to share discoveries for the benefit of science in general.[13]

The public were left with yet another promise that the discovery of the gene marker meant that Huntington's chorea could be eradicated (yes, eradicated) within a few generations. Reporting from a scientific conference, Germaine Greer had this to say:

'In one generation we could eliminate Huntington's chorea from the gene pool,' said one of the most distinguished gentlemen present. The only way this could be done would be by sampling genetic material from all foetuses conceived by people with the disease, and aborting all those that carried the marker for Huntington's. There was nothing in the great doctor's voice to indicate that he found this anything but a delightful prospect or that he anticipated any resistance from the Huntington's sufferers themselves. He had never heard of Woody Guthrie, who said, when he was in the final throes of the disease, that he thought he had had a good and useful life.[14]

The ethical dilemma most commonly recognised is that for people at risk from Huntington's chorea because of their family background, the genetic test may give them their lives back, or it may give them a death sentence. There is no treatment available for Huntington's chorea, just one of a growing list of conditions for which presymptomatic screening tests are available. Greer observed that about all the new genetics can offer a person suffering from Huntington's is to wish they had never been born.

Some people who know that Huntington's chorea is in their family history say they want to know if they carry the gene; others say they do not want to know. The personal dilemma has attracted notice, as well it might, but it seems to have got stuck at a point where it leaves the individual responsible for choosing testing rather than opening up the discussion to consider how the focus on genes-the-cause changes medical attitudes about certain diseases; how a legal apparatus is emerging about genetics; the stigma that may accompany a person diagnosed with a genetic condition; the pressures on people to embrace the gene technology solution, to name a few pressing problems.

Cystic Fibrosis

Cystic fibrosis is an incurable illness but one that many people live with. Treatment includes taking drugs for it, and daily, often painful, physiotherapy to decongest the lungs.

When the discovery of a cystic fibrosis gene was announced in September 1989, a friend said to me, 'What, this again?' It was the third time in the space of three years that such an announcement had been made. The 1989 gene was identified by a joint team from the University of Michigan and University of Toronto. We had already heard about such a discovery in 1987, but the 1987 gene merely localised the area of DNA. Although in 1989 much remained unclear about the nature of this discovery, it was said to offer wide screening potentials.[15]

The hunt for the cystic fibrosis gene was another shabby story of intense scientific rivalries for kudos and intellectual property rights, of ego-driven science, material greed, commercial interests, secrecy and alleged withholding of information. Leslie Roberts in *Science* said, 'Races in human genetics – such as those under way on Duchenne muscular dystrophy, Huntington's, and other diseases – typically these have elements of both cooperation and competition. But in the race for the cystic fibrosis gene, the competition has, at times, been extreme.'[16]

Cystic fibrosis is understood to be a genetic disease which requires both biological parents to pass on the condition. It affects mostly white populations and there are an estimated two million 'carriers' in Britain. (Tay-Sachs is another disease understood to be inherited in this way; only if both parents carry the gene and if the child inherits the gene from both parents will she or he have the condition. Tay-Sachs primarily affects descendants of central and eastern European Jews, although it may be found among members of any group.)

If so many people are carriers of cystic fibrosis, and if screening before a woman becomes pregnant is 'the most appropriate strategy', as some suggest, then a population genetic screening policy would need to be organised into medical practice and social practice. There are precedents of mass screening policies elsewhere, as in Canada for Tay-Sachs disease or in Italy for thalassaemia. Both examples are hailed as success stories which resulted in a decrease in the number of people born with the conditions after implementation of policy. But there are other stories with different endings, as in the following case.

Orchemenos is a Mediterranean village which became the setting for a medical experiment that went terribly wrong. In Orchemenos, there was a high incidence of sickle cell anaemia, a heritable illness which may be passed on to a daughter or son if both biological parents are 'carriers' of the disease.

Sickle cell anaemia is a blood disorder or, more specifically, an illness of the red blood cells. (Blood contains millions of round, disc-shaped red cells.) It is found in higher frequency in people of African, Afro-Caribbean, Asian and Mediterranean origins. A person with sickle cell anaemia is described as having two genes for abnormal haemoglobin (the oxygen-carrying protein in the blood) which gives their red blood cells a sickle shape, hence the name of the illness. A person with sickle cell trait is called a carrier. Such people are identified as having one abnormal and one normal gene for haemoglobin; they have some abnormal haemoglobin, but most of their red blood cells appear normal under a microscope. They do not have the illness sickle cell anaemia, and they cannot acquire it. However, people with sickle cell anaemia have periods of sharp and prolonged pain, anaemia and infections, called crises, when their red blood cells are less able to carry oxygen, due to any number of reasons including dehydration, overexertion, stress, and during and after pregnancy. Sickle cell anaemia is called a multi-system disorder, and hence children and adults who have it are also prone to other health problems, such as inflammation of the hands and feet. However, the crises and disability of the illness vary greatly in severity from person to person. There is no cure for sickle cell anaemia, and it is often fatal. But for many people there are various forms of treatment that may prevent or relieve symptoms, for example making sure to drink plenty of water, and the use of painkillers during a crisis.

In Orchemenos, the medical teams estimated that one in four of the villagers had sickle cell trait, that is, they were carriers of sickle cell anaemia. 'Scientists decided to help,' explained science writer Robin McKie in relating the story.[17] They decided to do so through the village marriage arrangement customs, to persuade carriers to marry non-carriers, and thus eventually free the village of sickle cell anaemia. 'Villagers were screened, carriers warned and the scientists departed.' The débâcle began. Carriers were shunned; they ended up marrying each other and the rates of sickle cell in newborn babies rose.

McKie concluded that the problem was education and counselling 'about the realities of genetics'. Otherwise, 'people had no way of making sense of medical advice.' He suggested that the lesson it held for Britain if it were to avoid a similar horror with new developments in gene technology was that, 'The Government must find ways of institutionalising genetic counselling throughout the country's general medical practices, family planning clinics and classrooms as a matter of urgency.'

I learned different lessons. This *was* a reality of genetics, and another example of the huge deficit of human understanding in the scientific 'solutions' of gene technology, and its potentially disastrous consequences. No matter how good-willed an individual doctor or scientist is in their wish to alleviate the suffering of human beings, this particular experiment was at best ill-considered, at worst a paternalistic attempt at scientific population planning. The implementation of an institutionalised genetic control of reproduction based on eugenic values made human beings and lives invisible. Human needs were interpreted from the abstract 'laws' of genetics. Human needs were sacrificed to the abstract idea.

There is something here to suggest that institutionalisation does not prevent people being stigmatising, but normalises it. What power or choice does an individual have in accepting or rejecting a genetic therapy if that is the only help on offer or when it is perceived as the best alternative? At the moment, many women who undergo genetic screening do decide to carry on with their pregnancies when cystic fibrosis is foretold by screening. For everyone who lives with it, the pain of having cystic fibrosis or sickle cell anaemia is not the only thing their lives are about.

Huntington's chorea, cystic fibrosis and sickle cell anaemia may well be illnesses which must be understood as inherited – but not *only* as inherited. First, it is ridiculous to reduce a condition to a singular hereditary notion of genes, not least considering that scientists admittedly have only the haziest ideas about how genes work in the body. Heredity is not simply the sum of one's genes. Hereditary does not simply mean genetic. Anyway, even if scientists 'knew more', it is no excuse for establishing any eugenicist policy. It does not mean medical care should inevitably become a policy of screening to eliminate a type of person. There are alternatives.

Consider, for instance, the rare biochemical disorder

phenylketonuria (PKU) which can be detected soon after birth and treated with diet. PKU is a condition in which the amino acid phenylalanine cannot be metabolised; it results in accumulation of phenylalanine in the blood which can produce brain damage and mental retardation. Early dietary restriction of foods containing phenylalanine prevents the damage. And even if elimination of PKU through screening was implemented, PKU would not be eradicated for all time. PKU-type conditions have been found in people who are not born with it, but who have been poisoned by excessively high intake of certain amino acids from synthetic foods.[18]

In addition, there has always been an alternative medical opinion which holds that ridding the 'human gene pool' of seemingly 'deleterious' genes might do more harm than good. 'Good' genes might be eliminated along with the 'bad' ones. Tay-Sachs is arguably inherited along with a positive trait, increased resistance to tuberculosis. Sickle cell trait is thought to confer increased resistance to malaria, which accounts for its prevalence among people originally from countries where malaria is endemic.

The above examples of Huntington's chorea, sickle cell trait and cystic fibrosis fit into 'classical' explanations of genetics and its laws of heredity. We now move on to conditions that cannot be explained with the mathematical 'laws' of classical genetics, but which have none the less been subject to much genetic theorising.

Schizophrenia

A research group in London's Middlesex Hospital under Dr Hugh Gurling identified the site of a 'genetic fault' correlating to schizophrenia in 1988. As other such discoveries, this one was prematurely based on small samples and resulted in rather hasty claims. Gurling's data suggested that much more schizophrenia is inherited than was once believed, which encouraged the thought that it should be possible to screen people and fetuses to see whether they are at risk of developing schizophrenia later in life. Another thought was that, having found the site of the gene on a particular chromosome, researchers now had a starting point to study the faulty brain chemistry believed by some to cause schizophrenia. This now seems unlikely.[19]

Within a few months, the situation proved more complicated. Another research team, lead by Dr Kenneth Kidd of Yale University Medical School in the USA, did not find the same genetic marker as Gurling had. Kidd did not doubt that a gene correlating to schizophrenia existed; he just predicted that a different marker would eventually be found in the family he was studying. It was far from clear what proportion of incidence of schizophrenia should be considered of the genetic variety. Furthermore, Gurling's family studies also showed that not everyone with the gene he identified developed schizophrenia. And despite the hype over the genetic link 'discovery', the genetic connection was as controversial as ever.

Controversy over the nature of schizophrenia is as old as the identification of schizophrenia itself. There is no general agreement about its cause, its diagnosis, its symptoms, its cure, or even whether it actually exists as such. Yet, 'one in every 200 people are diagnosed as having a schizophrenic illness at some point in their lives.'[20]

Psychiatrists were the first to define schizophrenia, but psychology, sociology and genetic science have their own claims to it as well. Theories have fallen into different camps, from the neurological (biological brain disorder) to the sociological. The renegade psychiatrist R D Laing believed that people expressing the behaviour called schizophrenic were making an intelligible response to an unliveable world. The psychologist and writer Dorothy Rowe explains that what psychiatrists call the delusions and hallucinations of schizophrenia are the fantasies of people who are giving meaning to their fearful world. Rowe suggests that while the person's reactions may appear inappropriate to the observer, they are appropriate and consistent to the person expressing them.

Schizophrenia remains difficult to diagnose with its spectrum of symptoms. It may be understood as showing

> dramatic disturbances of thought and feeling. People start to experience the world very differently. They may come to believe that their thoughts, feelings and actions are under the control of an external force ... They may experience visions, seeing, hearing or even smelling things that others can't ... They may be convinced of something for which there is no obvious justification.[21]

The mother of two sons suffering with schizophrenia described the illness as 'a personality disorder which doesn't affect the character ... His true self is intact. That's what keeps you going.'[22]

Mental illness is an area in which the limitations of Western medicine are glaring. This system thinks in terms of causes which generate illnesses, and this is also true of its approach to mental illness. There must be *some* cause, biological or psychological, which can be named. By contrast, psychiatrist Suman Fernando of the Chase Farm Hospital in the UK suggests, 'Biochemical and genetic influences might come to be better understood in dynamic balance or imbalance with each other, and with others – social, cultural, spiritual and cosmological.' Taking this approach, you do not have to identify some states and some people as 'pathological'; mental illness may be understood as disturbances of balance within individuals, families, societies.[23]

In spite of this insight, the field of schizophrenia is dominated by hereditarians who maintain that the underlying origin is located in the genes. What is the basis of the genetics of schizophrenia? The original studies which laid claim to it, and which remain a major influence on teachers and students of mental illness, were made by one Franz Kallman who published his data in the late 1940s and early 1950s. When J Richard Marshall went back to review Kallman's work, he found it riddled with distortion and fiction.[24]

To prove whether schizophrenia was genetically based, Kallman studied identical twins, since they inherit the same set of genes. Kallman's research mission was to identify one of a set of identical twins with schizophrenia, and then find out if the other twin also suffered from schizophrenia. He found the correlation between heredity and schizophrenia he sought, but his research methods and data are ambiguous and faulty. For example, Kallman himself made both the diagnoses of schizophrenia and the identification of twins as identical. By the most basic standard of good clinical research practice, these judgments should have come from independent sources. Schizophrenia, like other mental illnesses, is identified by behavioural symptoms, and is not a straightforward diagnosis. Also, adding to the problem of in-built bias, Kallman was a committed eugenicist who wrote 'prolifically about "mental illness", and to extol the virtues of eugenics and biological psychiatry, exhorting mental health professionals to apply genetic principles'. He called people with schizophrenia who did not require hospitalisation 'diseased trait-carriers'. When his methods and data were challenged, clarification was never forthcoming, neither from Kallman himself nor apologists of his work. Marshall concluded, 'Considering his emphasis on a scientific approach, and his staking of far-reaching eugenical proscriptions upon the results, such omissions amount to more than negligence.' Statistical studies may impress as scientific and impartial, but 'it should never be forgotten that however one juggles the figures with dazzling mathematical techniques, they can never be more accurate than the original observations upon which they are based.'

Kallman is not an isolated figure. A eugenicist perspective influenced much of the scientific – biological and psychiatric – work into mental disorders at the time. Many worked from a 'degeneracy' theory that vaguely interpreted mental defect, epilepsy, alcoholism, criminality and insanity as the result of 'poor heredity'.

Kallman's work cannot be dismissed as an unfortunate one-off incident. Marshall called it an example of how results could be distorted to fit an ideology, and which were then uncritically accepted by a scientific community which found the results supporting its own purposes. This, I feel, is a problem with so much gene hunting these days, including the efforts of so many researchers trying to find the genetic component of schizophrenia. Less than two years after the identification of a genetic basis for schizophrenia was announced, the British medical publication the *Lancet* carried a report of a study that suggested that the improved health of mothers was a factor in the substantial fall in the number of people admitted to hospital for schizophrenia since the 1960s. The researchers, who accepted that schizophrenia was partly genetic, concluded that the decline in the incidence of schizophrenia must be seen as due to changes in environmental conditions: 'The crucial factors were probably the improved nutrition and improved maternal health that came in with the Welfare State in the early 1940s.'[25]

There often is an over-riding sense of despair and concern among those who take care of loved ones with schizophrenia. But that sense of despair is not simply a result of the illness, but has much to do with the level of support and specialised care which can empower people suffering with the illness and their carers and friends. I heard a man suffering from schizophrenia call himself disadvantaged, which refuses the image of a broken self the way words such as 'defective' have come to connote.

Every kind of genetic research requires diagnostic judgments. Molecular gene analysis does, and so does psychological analysis through the study of inheritance patterns in families through observations of behaviour (studies which are still done).

Where does all of this leave the genetic hypothesis today? Oddly, well and strong. Today, many hereditarians believe that the genetic studies of diseases such as schizophrenia should be placed in the context of various environmental factors. That may appear more convincing, to say that it's a little bit hereditary and a little bit environmental. But environmental factors are mostly seen as influencing the working of the gene, often as a protection against or a 'trigger' for a genetic factor. Explanations still revolve around genes and solutions focus on control of the genes.

Consider the discipline of ecogenetics. 'Ecogenetics' may sound as if it redresses the lack of attention to environmental and social factors – ecological principles – but the fact is, the gene hypothesis has become more sophisticated to accommodate new knowledge. Our next step along thg gene route, which explains just how this is happening, is an examination of the theory of 'hypersusceptibility' which began with the claim that certain workplace injuries were the fault of an individual's genetic make-up.

The Hypersusceptible Worker

In 1986 an article appeared in the *New York Times Magazine* called 'The Total Gene Screen'.[26] It was about all the sorts of genetic testing which now exist to find individuals with, as it stated, 'tiny genetic flaws that foster deadly disease'. These genetic 'flaws' or 'anomalies', as they were called, explain why some people are more likely to develop cancer and other illnesses from exposure to chemicals, certain drugs and cigarette smoke.

It centred on a man called Ed Morgan who died of pleural mesothelioma, a type of cancer the only known cause of which is exposure to asbestos. He had worked as an insulation installer for 20 years. Most of the insulation was asbestos. It is well established that exposure to asbestos can cause different types of cancers and an occupational illness known as asbestosis, a disease of the lungs where inhalation of asbestos particles causes severe thickening and scarring of the lung tissue. The writer of the article did not mince words. He accepted the judgments of a new genetic diagnostic test, saying it was likely Ed Morgan 'was hereditarily more susceptible to asbestos . . . The problem was in his genes.' The article predicted that genetic screening for such variations is very likely to become commonplace in the workplace and beyond, into the community at large and the home. 'What is true in the workplace is true everywhere,' proclaimed 'The Total Gene Screen'. 'Many of the numerous serious diseases induced by certain pesticides, exhaust fumes and other pollutants, by drugs, cosmetics, charcoal-broiled and fatty foods and by radiation occur in those who are hereditarily vulnerable.' The message: Don't blame the food, chemical or radiation. Blame your genes.

With Ed Morgan's story, the article dramatised the resurgence of two closely aligned ideas, genetic screening of individuals and the theory of hereditary 'susceptibility' to illness. The theory of susceptibility and the practice of screening individuals were brought together by industrial toxicologists in the decade before gene splicing technology appeared. But genetic engineering opened doors to re-forming the theories, theories which remain as spurious as they always have been. Furthermore, the 'evidence' for the theory was built on unproven premises and sloppy reasoning.

Workplace genetic screening first appeared in the early 1960s when tests were available which could identify conditions such as sickle cell anaemia and some other genetically based metabolic (biochemical) differences among individuals. When occupational health specialists observed individual variation in individual responses to workplace environments (which is hardly surprising; not everyone exposed to the bubonic plague died from it), some offered that these differences in response might be genetic. They created a terminology to categorise people. The 'hypersusceptible' individual responds to 'low levels' of exposure and becomes ill, while a 'resistant' individual shows no response unless exposures were increased above an 'acceptable level'. The man credited for coupling the theory of 'acceptable levels' with the theory of the 'hyper-susceptible worker' was Herbert E Stokinger, advocate of workplace screening and chief toxicologist at the US National Institute of Occupational Safety and Health. Industrial toxicologists explain healthy and unhealthy reactions to environmental poisons (toxins) in terms of 'threshold levels' of exposure; that explanation assumes healthy individuals will suffer no ill effects below the threshold level. In fact, some individuals may be experiencing ill effects but to a lesser degree than those experiencing a full-blown, identifiable occupational disability. None the less, the theory of 'threshold levels' remains alive in the annals of industrial toxicology.

One of the first genetic characteristics for which employees were screened was sickle cell. Although, as I explained above, people with the sickle cell trait do not have the illness sickle cell anaemia, Stokinger labelled those with sickle cell trait 'hypersusceptibles'.

DuPont Chemical Company in Wilmington, Delaware selectively screened black employees for sickle cell trait on the unproven assumption that workers with it would get sick in an industrial environment. Black employees were the only group screened, although there were many employees from Italian backgrounds, among whom there is also a higher frequency of the sickle cell gene. The screening clearly had more to do with discrimination, racism and job exclusion than with the prevention of workplace injury.[27]

Job-related screening was challenged by trades unionists,

civil rights groups and women's groups, who recognised the problem from their own experience; women 'of child-bearing age' had often been excluded from jobs on the grounds that certain industrial environments posed risk of chromosomal damage to a foetus. The protesters argued that under US law employers are obliged to make the workplace safe for all workers and not just for some, augmenting their argument with reference to constitutional rights to privacy and equal opportunity, the right to work and the illegality of discrimination against people with handicaps.

Workplace screening remains an attractive proposition to industry, as became clear after the US Congressional Office of Technology Assessment carried out an anonymous survey in 1983. Of over 500 companies surveyed, of which 366 responded, 17 said they had used screening in the previous 12 years; five were still using it at the time; and 59 said they planned to initiate it, four within the following five years.

The 'hypersusceptible individual' theory, where all this started, was put forward by toxicologists who were hired either by industry or by government regulatory bodies who had an interest in the overall stability of the industry. (It reminds me of the Wellcome scientist's 'trigger theory' to explain serious adverse reactions in children from whooping cough injections [see chapter 6], thus locating the problem in the individual's biological make-up, and thus limiting industrial liability.)

The 'hypersusceptibility' theory is instructive for what it reveals about undercurrent eugenicist principles: eugenicist remedies are not limited to selective breeding. Toxicologists promoting the theory emphasised the benefits genetic screening would bring industries: health and safety would prove cheaper. Why? Because instead of being based in conditions of the plant or factory, it would be based in medicine. '. . . we are transferring control of worker exposure from engineering to medicine . . . with resulting shift in emphasis from general (process) control to control of the individual,' they boldly announced.[28] Restrict the person, not the substance.

This philosophy runs close to a principle of the racial hygiene movement, the German equivalent of eugenics. An historian of the period, Sheila Faith Weiss, explains that Wilhelm Schallmayer, the founding father of racial hygiene, believed that society's genetic composition was a genetic resource which had to be properly managed. The goal of 'racial hygiene' was to structure an industrial human society where medicine served the state, rather than the patient, through 'eugenic prevention'.[29]

The 'hypersusceptible worker' theory provided a model for explaining the ill effects from exposure to toxic chemicals and pollutants such as lead, petroleum fumes, formaldehyde-treated fabric and fumes emanating from plastic items. Instead of calling the effects what they are – chemical poisoning – it is called 'chemical sensitivity'. The notion 'sensitivity' is partly supported by measurements clinicians make of enzyme deficiencies in the blood of *some* (not all) people who get sick from exposure to certain toxic chemicals. Since enzymes are gene products, the supposition is that perhaps this is a gene problem, that these toxic reactions have a genetic determinant.

As someone who experienced chemical poisoning at work, I would protest against this unsatisfactory supposition and diagnosis ('chemical sensitivity'). It is just as likely that over-exposure to environmental poisons damages a person's ability to cope further with similar chemicals. Human beings should not be expected to inhale vast quantities of these substances and remain healthy. It could very well be that in some instances the blood tests are measuring a biochemical or genetic *effect*, similar to the chromosomal damage or genetic damage caused by exposure to nuclear radiation contamination, rather than a genetic cause.

Molecular biologists working in various fields, such as

cancer research, suggest that automated genetic profiling should be used to identify 'susceptible' individuals. How can anyone imagine that genetic profiling will do anything to improve health care when what can already be done to improve people's health is not being done? Even in a utopia, would a personal genetic profile be useful? If human geneticists are correct and all disease has a genetic component, what does that say? You might as well say the whole body is implicated in disease, or that if you did not have certain genes, you would not be a person.

The genetic hypothesis is beginning to sound like a circular argument. Consider the statement that there is a genetic link to old age. One professional in the field of Alzheimer's disease deduced, 'Depending on your genes, everyone will get it by the age of 120.'[30]

Or consider a cancer gene theory, the idea that we all carry the genetic predisposition for cancer and that something 'activates' the cancer-causing genes. If it is true, then all that the genetic hypothesis is saying is that human beings may get cancer or they may not, depending on a whole host of things. Taking bits of DNA out of the organism to study their cancer-causing potentials is not enough. What happens in the living organisms is important too. Where the living organism *lives* is important too. These aspects are fruitful avenues for understanding the multiple causes, origins and possible cures for cancer.

Genes-the-Burden

We are not laden with a 'genetic burden' or 'morbid genes'. We are weighed down by the concept of genetic burden. At the very least the genetic hypothesis is overworked, and not just in its pursuit of genes as a basic cause of the human condition. The gene theory is a package deal: genes are manipulated to explain disease, to diagnose disease and to cure disease. Genetic technology is both hammer and nail.

Or, to use the military metaphors of medical language, genes are both the weapon and the target. DNA tests are used to find 'faulty genes' in people, or to analyse the genetic profile of bacteria or viruses to diagnose infections. The symmetry is neat, as in the compact rendering, 'DNA probes set DNA to detect other DNA'.[31] Genes are further used to explain all sorts of conditions, not just ones associated with a 'faulty gene'.

The new genetics is changing the meaning of cause and prevention. Genes become the cause of an ever-expanding list of human conditions. Genetic technology becomes the solution. Genetic intervention becomes prevention. This is a shift away from previous ideas that to prevent ill health was to forestall the development of illness or disability through social means such as adequate nutrition, housing, unpolluted air and water and health education.

This is not to deny the significance of our bodies – of physiological processes, of physical or biochemical explanations of things. There *are* biochemical explanations of things. Malnutrition can be understood as the lack of chemical nutrients (vitamins, minerals, protein, carbohydrates and fats) which are found in food. Though the biochemistry of nutrition may document the conditions, it does not explain why people are starving. Nor has a knowledge of nutrients made scientists able to create exemplary synthetic food (although they have tried). Similarly, the chemical substances known as genes have an important biological function. But they are not the keys to human health and welfare. We are not machines and genes do not act in isolation.

We return to the question, why the relentless pursuit of the 'new genetics in clinical practice'? And we return to some familiar answers. It satisfies heroic science, it is fashionable and hence funded, it is lucrative with its 'gene probes' and tests which can be marketed (diagnostic gene probes alone have an estimated market worth of at least US$500 million by 1992[32]), and, as the theory of the hypersusceptible worker exemplified, it may serve industrial interests and lingering eugenicist ideals. All of these elements of the new genetics in medicine are shrouded by the over-optimistic belief in the curative value of genetics and an implicit faith in science.

Perhaps not all races for genes are as exemplary in their notoriety as the ones inspected in this chapter, but attempts to find and classify human genes have not come out of thin air. What I have been concerned with in all the above examples is to suggest that something far more serious is happening than just individual examples of misapplied science.

In making reference to the history of eugenics and especially to Nazi political philosophy, I do not want to give the impression that what is happening today is exactly the same. That would diminish the exploitations and the unspeakable atrocities and repressions of that time, the use of human beings as fodder for population controllers, the attempted genocide. And it would do us little service in seeing the important differences, in analysing where medical genetics is going today, and in appraising its presumptions and premises, its goals, and the means to reach those goals.

But to ignore connections is equally diminishing. Dr Jeanne Stellman of the Women's Occupational Health Resource Center at Columbia University commented on the incidence of workplace screening:

> I am the child of a survivor of the Holocaust ... and many millions of others suffered very directly and convincingly the results of policies based on dubious genetic traits ... While I don't see gas ovens being built in the United States, I do see that many of the premises which led to that terrible time are again beginning to surface here.[33]

Today's biological theories of the human condition, couched in terms of genes, are affecting personal expectations, and legal, medical and workplace controls on people's lives. How far is genetics to be used to define fitness to work or breed, what proper sexual and reproductive behaviour should be, and which kind of people there should be?

When genetic control becomes equated with medicine, when eugenics becomes institutionalised, we are in trouble. What kinds of medicine there should be is a political question. Eugenic modes of medicine have always been haunted by their normalising role. Who is normal? Which genes are normal?

8 Reproductive Engineering

> Biotechnology will have a number of profound impacts
> upon reproductive technology, ranging from the pre-
> vention of pregnancy to manipulating the genetic make-
> up of children.
> Sue Meek[1]

The industrialised world has taken on a distinctly genetic
flavour. Today it is almost impossible to think about health
and reproduction – whether of plants, animals, bacteria,
ourselves or any other living thing – without thinking
about genes. The genetic connection in human reproduction
is nowhere stronger than in the practices of reproductive
technology, a term which came into its own with the arrival
of 'test-tube' babies, but which has come to mean any of the
science-based or medically controlled or commercially
controlled interventions in female and male fertility. It
spans contraceptives and sterilisation, donor insemination,
fertility drugs, the test-tube baby methods of *in vitro*
fertilisation (IVF) and related procedures, surrogacy, genetic
screening, performing operations on foetuses during
pregnancy, using foetal tissue acquired from women who
have abortions in medical procedures, growing embryos
and foetuses in the laboratory and experimenting on them,
freezing and banking of embryos and gametes (eggs and

sperm), the prospect of gene therapy (genetic engineering) in embryos and foetuses, and variations on these themes.

Genes and Reproductive Technology

The priority in medicine and in science given to the genetic dimensions of reproduction was inescapable in a television programme called 'Genesis', aired 13 January 1986, in the BBC's leading science, medicine and technology series, *Horizon*. The subject of the programme was embryology, the scientific discipline which brought about the methods of *in vitro* fertilisation and the research on women's eggs and human embryos which went before and continued after the introduction of such methods. 'Genesis' was mainly about genetics. It showed a chimeric mouse and clones. It compared the lives of cancer cells and embryo cells, suggesting they were similar but that the cancer cells differed in that they did not 'turn off', as did cells during the growth stages of embryo development. It showed a horse giving birth to a zebra, the zebra embryo having been placed in the mare's womb. It showed an example of a human baby born from a frozen human embryo. Biologists appeared and talked about the central problem in the study of embryo development: how does a single cell (the female egg cell) become a mature creature? By and large, they said, this occurs through the effects of genes on growth and metabolism. An important principle of embryo development is that the chains of events which lead to different kinds of tissues and organs in the growing organism are controlled by genes. Specific genes may determine which of two alternative paths of development may be realised. One scientist, a developmental biologist, explained how he managed to create a fruit fly with its head located where its bottom should have been.

'Embryology has become genocentric,' declared scientist Michael Ashburner. 'The past twenty years have seen the integration of genetics into embryology, the realization that developmental processes can only be understood in terms of their underlying genetic bases. This has led to a shift from what I will call "soft" genetics to "hard" genetics.'[2]

I cannot stress enough the importance of this interest in the genetics of embryo development, or should I say, *pregnancy*; or the link that is made with other areas of medical genetics and the implications for clinical practices. One of the suggested reasons for mapping and classifying all the human genes is that, in total, the human genome 'contains all the instructions needed to go from a fertilised egg to a complete human being'.[3] Knowledge of the genetic mechanisms involved is of great research interest. The direct line to 'practical application' was explained as follows by medical correspondent John Illman in the *Guardian* of 1 March 1989:

- The 'new genetics in medicine' offers a 'comprehensive assault against disease'. It will not only affect the kinds of medical treatment we are given, but what we eat, what jobs we hold, how we live – depending on our known predispositions to disease, knowledge of which the new genetics will provide.
- Today, 'test-tube' baby technology offers another avenue for genetic screening: on the fertilised egg (embryo) in the laboratory dish.
- In the very near future, another method for embryo screening will be available, namely flushing out the early embryo from a woman's womb so that it can be tested, and then returned to the womb or discarded depending on the test results.
- There are two keys to the success of these objectives. One is the present project to map and sequence all the human genes, which should be completed by the early twenty-first century. It will enable the identification of some 4000 listed genetic disorders. The other is a technique called DNA amplification, which allows

detection of genetic material in just one or a very few cells; a single cell can be removed from a human embryo at an early stage (two or three days after fertilisation in the laboratory dish).

At the time Illman wrote, some 400 IVF embryos had been tested by embryo researchers experimenting with the method in British laboratories.

In summary, between the desire to explore the genetic mechanisms of embryo development (what happens in pregnancy) and the motivations to expand 'the new genetics' to reproductive medicine, there is a major interest in laboratory and clinical research on women's eggs, fertilised eggs, embryos and foetuses – whether in a 'test-tube' or in a woman's body. Sometimes that interest sounds exaggerated but none the less 'logical', as in the predictions of the director of the Centre for Cellular and Molecular Biology in Hyderabad, India, who wrote in 1985:

The *real* age of test-tube babies might dawn too, so that the *entire* process of not only fertilisation but of subsequent development [of the foetus] might be carried out outside the body of the female of the species . . . [and] . . . Much earlier than we think would come the technique of changing the genetic make-up of man by introducing foreign genes or by replacing cells in the egg at various stages of development. That it is possible to do so, has already been demonstrated with experimental animals. One can predict the appearance, on the international business scene, of gene banks which would supply a package of genes (both synthetic or natural as one might like or as might be convenient) – coding for a required trait, which could be used to introduce a desired modification in the progeny.[4]

The first human egg bank was planned to start in Singapore in 1988.

Research interests abound. So do clinical interests. Both can be understood as social interests. We are meant to understand the social interests as in the interests of society, individuals, groups, women. Are they? You can pick and choose which genetic interventions you want. Or can you?

Reproductive Biotechnology

It has been sobering for feminists like myself to experience how invisible women have been in so many of the scientific and political analyses of the biotechnology revolution. While critical questions have been raised regarding the risks of reproductive engineering in micro-organisms, plants and animals, or questions regarding the power and control of transnational corporations in other areas of biotechnology, such questions have not been raised in connection with the biotechnology of human (read: women's) reproduction. But similarities there are. Similar economic and scientific (that is, genetic) priorities influence the path along which reproductive medicine is travelling, as in the following examples.

Commercially, reproductive technologies satisfy the biotechnology industry. Pharmaceutical companies profit from them. Many of these companies, which have expanded their interests to include genetically engineered crop seeds, are suppliers of fertility and contraceptive drugs. They often fund research in the area of reproductive technology, much of which utilises their products. They support scientific meetings and conferences on human reproduction; these scientific meetings dwell on the technological approaches to medicine. The pharmaceutical giant Serono bought Bourn Hall Clinic, the biggest IVF centre in Britain. Surely these relationships present conflicts of interest. (What is good from the point of view of a drug company's accountants may not be so good for the person receiving a drug treatment.)

In the United States, commercial surrogacy arrangements are organised by entrepreneurs like the lawyer, Noel Keane. While surrogacy, where a woman bears a child for another person or couple, is not by definition reproductive technology – you don't need modern scientific knowledge or know-how to accomplish it – it has come under the umbrella of reproductive technology for other reasons. Firstly it emerged around the same time as 'new' reproductive technologies like IVF which were being used for infertility; and it fulfilled a somewhat similar role, albeit through a very different means. (Further, surrogacy is brought to IVF in 'IVF-assisted surrogacy' where the 'surrogate' mother receives an embryo created from the egg and sperm of the couple who will receive the child when it is born.)

The proprietary philosophy extends not only to women's capacity to bear a child, but to women's eggs, as in the paid recruited egg donation programme begun in California in 1990; women donors were paid $1500 if they took fertility drugs to stimulate their ovaries, and if they produced eggs for extraction. If no eggs were produced, the volunteer received $500. One doctor who used eggs acquired thus said the payment was not for the woman's ova (eggs), but for her services. It was not the first such programme, nor is the United States the only place it happens. As for other avenues of acquiring eggs, Dr Charanjit Babra, who does fertility research on baboons in Kenya and is a consultant on anti-fertility vaccine contraceptives for the World Health Organisation, admitted that just as researchers found it easier to get women's placentae after abortions in India than in Kenya, he believed there would be 'no problem' getting women's eggs for experimentation by plugging into Indian IVF programmes in future. This work, he suggested, would be impossible or at least more difficult in the UK where embryo research is allowed but only up to 14 days after fertilisation.[5]

New vistas of research in human reproduction are only

possible with the availability of human body fluids, cells and tissues. It is not so much that other living things – bacteria, maize, sheep and so on – are being used as the raw materials of reproductive engineering. All living things are, including human beings whose cells, tissues and genes are as important to research and clinical practice as are the cells, tissues and genes of other living things to animal and plant research. And of course, techniques used on animals to manipulate their reproduction are also used on people. For biological research in general and human reproduction in particular, parts of women's bodies are of central importance: eggs, embryos, placenta cells, foetuses, fallopian tubes, cervical fluids. Many governments have taken it on themselves to discuss 'cell line management', pondering, for instance, if women have any proprietary 'rights' to foetal cells and cell lines created from foetuses; this, too, like surrogacy or payment for women's ova (eggs), can be called the commercialisation of motherhood.

The similarities between the 'management' of animal and human reproduction are not hidden; they are very clear, as illustrated by *Manipulating Reproduction*, a book in the series *Reproduction in Mammals*, edited by research biologists Colin R Austin and Roger V Short in 1986. First, a veterinarian talks about the use of reproductive technologies or 'artificial reproduction' to increase productivity in farm animals, then the book goes on to discuss their uses in human reproduction: contraception, fertility control, infertility control, genetic screening, genetic engineering and, lastly, the 'barriers to population control' and how to overcome them through science-based reproductive technology. (When critics on this subject use language such as 'manipulating reproduction' to describe the same thing, we are said to be overly emotive and scare-mongering. Yet a group of eminent scientists may write a book of that title with every chapter but one about human – often women's – reproduction, and the connotation is acceptable.)

In and of itself, the practice of controlling our reproductive lives is welcome, not a problem as such. But control of our own reproduction is not necessarily equivalent to scientific control of reproduction. Perhaps the words of an ethicist whom we will meet again later, Joseph Fletcher, who believes in genetic control of reproduction, makes the point. He wrote, 'In the twentieth century there has been a conversion of the scientific motive from "just to know" to "how to control with what we know".'[6]

Social Repercussions

These developments are dramatic and raise many questions about the role and authority of scientists and entrepreneurs in steering the course of reproduction. The question that preoccupied scientists and ethicists contemplating this future in the 1960s and 1970s was not in particular 'How can we help women?' but rather, 'Should man play god?' They decided to allow themselves to play god. The ethical discussion was severely limited to the fear that the new 'artificial conception' technology could disrupt the ties that bind society, namely the institution of the family. This in itself should set off alarm bells as to the sexual politics at stake. From the point of view of women's health, sexuality and reproduction, the crucial question is, How do these approaches affect our lives, and what is the ideology that allows them to happen?

Lack of critical attention may seem understandable at first. Who can argue with the aim of curing diseases, making healthy babies, preventing unwanted pregnancies, helping women and men who are infertile to have a wanted child? On the other hand, the same basic questions as arise in other areas of genetic technology also apply here, and some crucial other ones as well. What does it mean that so much in terms of time and resources is spent on these avenues of genetic control? What are the objectives of these technologies? Why does medical science continue along the path of high-tech fertility intervention even when particular methods are found to be dangerous and ineffective? Equally important, as feminists pointed out in the early days and as the World Health Organisation eventually concluded, IVF patients – women – are experimental subjects of these methods and of similar methods – whether the research is clinically oriented or not; whether it is directly involved with women's reproduction or is not. The research on human eggs and embryos begins with women and involves women who undergo the interventions necessary for such scientific and social experimentation to happen.[7]

In Australia, the media has for a long time taken seriously feminist thoughts on reproductive technologies. The Green Parties in Europe have taken account of feminist concerns; the European coalition of Greens earmarked human reproductive technologies as one of four key political campaign issues on biotechnology. The other three were the genetically engineered hormone BST, the risks of deliberately releasing genetically modified organisms into the environment, and the patenting of living things. Some six years earlier, in 1982, the Sydney Women's Electoral Lobby called for a moratorium on reproductive engineering programmes.

Certainly, there are many issues at stake here, and there is no general agreement among feminists or anyone about their implications. When new methods like IVF appeared, they brought infertility to the foreground. We finally realised that up until then, infertility had been generally ignored, denied, or dismissed everywhere, including among feminists, women's health movements and reproductive rights movements in Western countries. So much attention had been placed on women's right to choose whether or not to have children, but none on the lives of women who found themselves unable to have children. For infertility treatment, new technology has resulted in some women having babies, although it has not served the needs of most women and men seeking medical help for infertility. It has also raised all sorts of questions about the politics of reproduction, sexuality and motherhood.

In some ways, the questions that arise are not all that unfamiliar to feminists: questions about medical and social controls over women's bodies and lives (for example, the reassertion of patriarchal norms by the worldwide trend to limit eligibility to 'artificial conception methods' such as insemination or IVF to married women or to women living in heterosexual relationships with men). In some ways, however, the questions which arise are of a different order of magnitude. For example, up until now the biological mother was the woman who gave birth to the child. That was certain. Now that women's ova can be removed, an embryo created outside the female body, and that embryo placed into another woman who may become pregnant and give birth to the child, for the first time in human history the question arises, 'Who is the biological mother, the woman whose egg was used or the woman who carried the pregnancy, laboured and gave birth?' Another unprecedented example is the capability to change the genetic constitution of a human being by genetic engineering of embryos and foetuses.

One thing is clear. Reproductive technology is affecting both the medical and legal worlds, and the biological and social aspects of reproduction. The social experiment that is reproductive engineering is having repercussions on women's lives, and not just in our reproductive lives. If reproduction is a major arena of sexual politics, so is reproductive technology. It is the marriage of reproductive and genetic technologies which I am interested in exploring further.

Embryo Screening

From the beginning, genetics has been high on the agenda

where growing women's eggs and embryos under laboratory conditions is concerned. That interest, as I
described above, takes many forms: for research to gain
knowledge about embryo development, and for interventions in the reproductive process. The intervention I
wish to talk about now is genetic screening of IVF embryos
before a decision is made whether or not to place them into
the woman's womb.

IVF was first a technique of animal embryology, to study
ova (eggs), fertilisation and early embryo processes. The
idea of IVF did not begin in the infertility clinic, but when a
few animal embryologists decided that they would like to do
similar research with women's ova, and found gynaecologists who were interested in collaborating, their initial
stated aim was to find treatments for infertility. (Although,
as many analysts have pointed out, IVF does not treat a
person's infertility, but bypasses the cause of it. This may
seem a small point, but taken in the whole context of
infertility services, it is an extremely important one.)[8]

IVF treatment always entails interventions for the
woman, whether the cause of infertility is diagnosed as the
woman's or the man's (sometimes IVF is prescribed when
the man's sperm is diagnosed as the cause of infertility).
Usually, the woman is administered fertility drugs and
hormones to stimulate her ovaries to produce many eggs
during the one cycle instead of the usual one egg per cycle;
and to control other processes such as the time of ovulation.
Eggs must be surgically removed. If one or more embryos
are created and are judged normal, a certain number are
placed into the woman's womb. Inserting more than one
increases the chances of success at establishing a pregnancy;
inserting too many increases the risk of multiple pregnancies and births.

As a clinical practice, embryo screening took longer in
coming than the use of IVF for infertility treatment.
Researchers needed time to develop specialised diagnostic
tests and methods such as DNA amplification. They needed

access to human embryos for some of that research; the experimental embryos have mostly been provided within IVF infertility programmes – the 'spare' embryos made from all the eggs collected but which were not or could not be used for infertility treatment. In addition, in Britain at one time, some women were asked if they would donate eggs to research when they were seeking sterilisation operations. (Some women were offered free private sector sterilisations in exchange for their agreement. When news of the practice became public, it was banned by the Voluntary Licensing Authority, the body set up by the Medical Research Council and Royal College of Obstetricians and Gynaecologists to regulate IVF research and practice in anticipation of a government-controlled statutory authority.)

Professor Robert Winston, an infertility specialist at the Hammersmith Hospital in London, was among the first clinical researchers in Britain to develop embryo screening tests. His was the first clinic in England where women became pregnant after embryo screening; the newspapers quietly reported these pregnancies in 1990. By 1991, it was apparent that the announcement made back in 1987 – namely, that IVF embryo testing on screen for genetic illness was ready to be put into practice – was wrong. At scientific conferences, one could find practitioners who were sceptical about the present and future technical success of embryo diagnostic tests. Still, the idea has taken hold. (It reminds me of the present state of pre-employment screening as discussed in the previous chapter. Although presently the use of both forms of genetic screening are rare, and although either or both may never be applied widely, the idea is potent.)

Today, the growing acceptance of the idea of embryo screening – in medical and scientific circles, by government committees and ethics committees pondering the worth of IVF – is most perplexing. As many researchers, analysts and clinicians, first among them feminists, have pointed out, the IVF methods have great problems riding in their wake, including intolerably low success rates, physical risks to the women and to any babies born, and unknown long-term effects, in clinical and in social terms.

First, although thousands of babies have been born all over the world from 'test-tube' conception since 1978 when the first 'test-tube' baby was born, the success rate world-wide remains so dismally low (on average less than 10 per cent in clinics in Britain, for example) that one Australian government report went so far as to say that there is no empirical evidence to say that IVF has a higher success rate than no treatment at all for infertility. Most women who enter such programmes do not go home with a baby. The situation is extremely troubling, not only because so many women go through so much and do not get pregnant, but also because many people – the women and men involved – leave the programmes emotionally shattered. This is a failing of the 'success story syndrome', which measures success solely as getting a baby; it leaves many of the women and men seeking help for infertility without the kind of services which help them get through the crisis of infertility, to heal and to thrive.[9]

Second, although any physical intervention carries risks, the risks of the IVF interventions to the women and to any babies born have gone largely unnoticed in government and medical science circles. (These establishments have instead focused attention on the embryo.) For example, the various fertility drugs and hormones prescribed to women on infertility programmes have not been acknowledged. These drugs are known to carry risks to the women who take them, including dangerous effects such as hyperstimulation of the ovaries, ovarian cysts and cases of ovarian cancer. IVF pregnancies carry greater risks of ectopic pregnancies, miscarriage, pre-term delivery and health problems for those babies born. When fertility drugs are used or when multiple IVF embryos are inserted into a woman's womb, a higher number of multiple pregnancies and births results; often women are having twins, triplets and quadruplets.

Higher order multiple pregnancy and births are physically and emotionally difficult for women, and are hard on any babies which survive, and on all the parents and carers of the children. There have been situations where, when five or more embryos were placed in the woman's womb and have implanted, selective abortion of some of those foetuses (called pregnancy reduction) has had to be performed. All of these situations present health complications for the women and the children who survive. Some women suffer complications from the egg removal procedures as well. There are a few reports of life-threatening effects from the action of hyperstimulation drugs on women's ovaries, and a few cases of deaths associated with such treatment.[10]

All of this is now known and much was predictable. It is perplexing that in the light of all of the problems with IVF alone and with the use of fertility drugs which predate IVF – that problems which become more troubling as time goes on – instead of reassessing the worth of egg removal and other IVF practices, their use is expanding to include genetic screening of embryos. Why promote IVF embryo screening when the interventions which a woman must undergo to achieve IVF pregnancy are so fraught with problems: known and unknown health risks to women and any babies born, the incredibly low success rates, not to mention unprecedented social reverberations which have not been adequately considered (as in the question, 'Who is the biological mother?')? Is genetic screening deemed so helpful, so important, that the dangers and other considerations do not arise? In future, will women and men who are carriers for, say, cystic fibrosis be expected to go through IVF to have a baby? Many advocates of embryo screening say that it gives women worried about having a child with an inherited condition a better choice than, say, amniocentesis which cannot be carried out until the sixteenth week or later in pregnancy. But all along, the difficulties for women and the complications of IVF have been considerably ignored or forgotten. In addition, embryo screening itself may pose long-term risks. It may take twenty years to discover if a child born has suffered some ill effect when, say, a cell has been removed from it as an embryo. With these considerations in mind, it is exasperating to hear advocates of new reproduction technologies claim the high moral ground – that they are interested in alleviating suffering and that the answers lie in more such research and more genetic control. Where else do possible answers lie?

Genetic Control

Embryo screening is just one example of the relatively new possibilities for genetic interventions in reproduction. Another, though as yet much more controversial and untried as far as we know, is gene manipulation or 'gene therapy' of embryos or foetuses (embryo is the term used for the earlier stages of pregnancy).

In *Manipulating Reproduction*, one article ventured into a discussion of 'Reproductive options, present and future'. Its author, leading embryologist Anne McLaren, considered some already existing medical care for children born with 'birth defects', but found these lacking; genetic engineering was a desirable goal, and this could be accomplished through manipulating fertilised eggs in the laboratory. In her words:

> From a long-term point of view, however, it is undesirable merely to provide a cellular crutch that will enable genetically defective embryos to survive and possibly reproduce. Recombinant DNA techniques coupled with methods of gene transfer may eventually enable us to replace defective genes with their normal counterparts. If this could be done in the fertilised egg, possibly with the aid of nuclear transfer, genetically defective individuals would be able to reproduce without fear of transmitting their abnormality to future generations.

Genetic defects can be cured only by genetic engineering.[11]

This is quite a sweeping claim about what is desirable and
undesirable. It does not intimate any social complication,
for instance that women and any babies born thus are the
subjects of such social experiments, or that the implications
for people with the so-called defects are ill considered. I
suppose if one starts with the premise that genes are a
problem, genetic engineering must sound logical, socially
useful and desirable; women should welcome any of these
interventions as a reproductive choice. But that presumption is notably inconsiderate of the lives of women and of
people born with disabilities associated with genes, and
certainly ignores the fact that women are the main subjects
of genetic screening and genetic control, and that genetic
control of reproduction primarily occurs through women's
reproductive lives.

The scientific logic of human engineering is often comp-
lacent about the ethics of genetic control. When the scale of
genetic engineering became apparent in the early 1970s,
when it was seen to be a monumental act of defiance against
nature by those doing it, ethicists and philosophers became
involved in the in-house scientific debates over it. In 1974 a
book called *The Ethics of Genetic Control: Ending Reproductive
Roulette* by the theologian and professor of medical ethics
Joseph Fletcher proposed that biological selection and
control are ethical imperatives. (Reproductive roulette is
human reproduction which is not scientifically genetically
controlled.) He believed then that the state has an interest
in compelling the non-birth of genetically handicapped
individuals. He supported biological selection and biological
control of human reproduction to the point that it should be
compulsory if necessary. As he put it: 'There are too many
who do not control their lives out of moral concern . . .
Large families and a pious disregard of genetic counselling,
like refusing to undergo vaccinations until it is made a

matter of police enforcement, show how the common welfare often has to be safeguarded by compulsory control.'[12]

The mingling of the language and medical ethos of infectious disease and 'genetic disease' is deliberate. Fletcher said that there are more Typhoid Marys carrying around genetic disease than infectious disease. Admitting that compulsory controls on reproduction were as yet unconstitutional in the United States, he protested, 'Here, as in so many other ways, the law lags behind the ethics of modern medicine and public health knowledge.'

An influential intellectual, Fletcher hits a nerve that recurs throughout the medical ethics debates on genetics. His general views on compulsory genetic counselling are not necessarily shared by other professionals in positions of power. But the question of compulsory screening is a viable one. For example, an international survey of medical geneticists, grouped by country, presented to participants at the seventh International Congress of Human Genetics held in Berlin in 1986, broached the subject of mandatory carrier screening for cystic fibrosis. Although a majority of countries represented asserted a strong consensus for voluntary, not mandatory, screening, a remarkable number were not so certain.[13]

Fletcher and those who share his view have adopted a thesis that reproductive rights are not absolute and those who are identified as 'carriers' of genetic disease should be restricted from reproducing. Nancy Wexler, a social researcher well versed in the field, is worried about such attitudes. If the state should have such an authority, 'Should gangs of medical thugs rove maternity wards looking for erring parents to sterilize?'[14]

Fletcher's is not a woman-respecting ethic. Instead of forging an ethics in terms of the good of all people – women included – it defines it in terms of some scientific or eugenic parameter that requires certain groups – and all women as a class – to subordinate ourselves to the population control

(non) ethic. It seems that whether the ethics of reproductive engineering are moulded to suit traditional family
norms or 'alternative' ones, they remain devoted to
eugenics.

Fletcher captured the essence of a narrowly science-
defined ethic when he wrote:

> Sanitation and preventive medicine have already raised
> the quantity of people (with some serious dangers
> resulting as undesired side effects); now, as a new turn in
> our affairs, we are at last in a position biologically to
> increase the *quality* of the babies we make. All of this adds
> up to what Joshua Lederberg calls 'orthobiosis' – setting
> things right by applying the life sciences.[15]

Lederberg, who studied medicine, was awarded a Nobel
Prize for his work in bacterial genetics.

The science of the scientist (Lederberg) has influenced
the ethicist (Fletcher). Fletcher explains that Joshua
Lederberg talked about orthobiosis in his presentation at
the Nobel Symposium in Stockholm in September 1969,
'Orthobiosis, the Perfection of Man', where he lamented
how people wanted to put brakes on science. 'The suppression of knowledge seems to me unthinkable . . . How can
the ignorant know what they should not know?'[16] He was
speaking of research in human reproduction.

'Wrongful Life'

Who is being ignorant? A legal category called 'wrongful
life' emerged in the 1960s in the US courts in regard to
illegitimacy. 'Wrongful life' suits allow a person to bring
parents and/or physicians to court for giving him or her life.
At issue in wrongful life cases is the plaintiff's contention
that because of extreme physical or emotional anguish, the
plaintiff should never have been born. Something should

have been done to prevent it. Today that may mean any of the numbers of medical interventions such as genetic screening either to predict a 'wrongful life' before conception or prevent it coming to term after conception.

Being born 'illegitimate' has long been stigmatised in patriarchal societies. Being born a female, or with a particular disability, may also stigmatise a human being. But the principle behind 'wrongful life', that there are lives 'not worth living', must now also be tied for ever to recent history: the racial purity ideals of the eugenics movement. Of course, court cases of 'wrongful life' are initiated by the individual, not the state. But the *legal* acceptance of the principle is significant. The courts in the United States have accepted a few cases of wrongful life, but to my knowledge, there has been no judgment made in favour of it. Joseph Fletcher felt that 'the principle is taking form, inevitably'. Nancy Wexler contends that allowing such court cases encourages parents to avail themselves of genetic services; such a legal concept puts stresses on parental responsibility regarding so-called genetic insults. By the 1980s in the US there were already a number of cases of women being forced to undergo Caesarean section births against their will by their doctors who sought and received court injunctions to carry out the procedure on the basis of their medical judgment that the foetus would otherwise be endangered.[17] The doctors call it taking a foetal precaution; the doctors claim that in these cases vaginal delivery would jeopardise the health or life of the child. In the United States the foetus is recognised as having legal rights from 24 weeks onward.

There are other cases, too, in which women have been restrained or taken to trial for failing to behave in approved ways during pregnancy – for example, failing to see a doctor in time or taking illegal drugs. How close is the trend coming to mandating for other sorts of medical interventions to 'prevent handicap', especially considering all the attention lavished on 'the embryo' and 'the foetus', while women remain far less visible as the human subject of our own fertility?

With compulsory vaccination and immunisation, you may remember that, despite varied opinions on the matter, it is medical heresy to question their effectiveness and safety today (unless it is used to rationalise more research on a better and safer vaccine, that is). When a strong-willed woman like Walene James dared to challenge the medical profession, the state and medical establishment tried to quash her resistance. And we are only talking about one woman's wish to refuse the compulsory vaccination of one little boy, her grandson. Was she a menace to society? What were they afraid of? If genetic screening becomes so much a part of the medical ethos of 'prevention', how much choice will a person really have regarding it? Even if we never come to the stage where genetic controls are made compulsory by law, social pressures to accept such interventions or the behaviour they suggest may still be great.

Motivations behind pursuing a particular line of research and medical practice are manifold; as are the motivations of doctors in the clinic, and the users of the medical care system. When doctors I meet say they see genetic screening being offered to the individual woman as different from population screening, I accept that they are sincerely motivated by a desire to help at a particular moment in time. The motivation is not irrelevant, but neither are the biases, the limitations, the social repercussions, the lack of imagination, or the power to coerce.

The fact that influential ethicists or medical scientists can explicitly or implicitly advocate 'manipulating reproduction' for genetic control as if it is of proven worth or is an ethical imperative is a testament to the attraction of the idea. Yet it is not the only way to go about arranging medical science research and services; it is not the only way to deal with infertility and human reproduction. To say this is nothing new. The history of science is a history of choices made and choices not made.

Sex Selection and the Maldevelopment Model

The social reverberations of reproduction intervention technologies for women all over the world can be understood better if you see how scientific developments can fit within the industrial model of growth and (mal)development. Reproduction engineering is logical if you are into creative economics. There is the profit to be made from the drugs, medical and laboratory supplies and specialised equipment necessary to the technology. There are the cost-benefits analyses which have been made to argue that genetic screening and selective abortion are cheaper than providing social services and specialised care for people with disabilities. But quite apart from these, there is a commodities angle – having babies is an enterprise like any other. The industrial view of plants, micro-organisms and animals can be applied to our own human reproduction. 'Children are increasingly being seen as products, and the new technology of reproduction, including the sale of reproductive material and services and especially prenatal diagnosis and selective abortion, encourage this commodification of the foetus,' wrote Barbara Katz Rothman, a researcher in the US who has studied women's experiences of prenatal screening. Rothman, like many other feminists writing in the area, sees a thread of the technology as 'treating people and parts of people – our organs, blood, energy – as commodities. When we talk about the buying and selling of blood, the banking of sperm, the costs of hiring a surrogate mother, we are talking about bodies as commodities.' Another writer on the subject, Ellen Hopkins, points out that women have always had concerns about the health of their babies, but now outlets for these worries are different from those of previous generations. In an advanced consumer society, 'perfect' designer children can become a social pressure.[18]

In twenty-first-century reproduction, you will be able to choose the sex of your baby, the book *Manipulating*

Reproduction reflected in the same breath as it considered how farmers have dreamed for many years of a 'method that would ensure that progeny were of the desired sex' for the economic rewards it would bring.[19]

Sex is a genetic trait, scientifically speaking. Biological sex is determined by the X and Y chromosome pattern in the cells. An XX pair of chromosomes identifies a female; an XY pair identifies a male. But variations on these two themes are also found, such as the XXX or XYY patterns which some people show. Also, there are examples where the person is male to look at, but has an XX (female) chromosome pattern.

With the use of 'gene probes', it is apparently possible to identify the male-bearing Y chromosomes in embryos. Such tests have been developed by IVF teams at the University of Edinburgh in Scotland and the Hammersmith Hospital in England, among other places. Other new diagnostic tests for sex are also being developed. In 1990 researchers at the Imperial Cancer Research Fund announced that they had identified a gene on the human Y chromosome which possesses the characteristics of a 'switch' guiding how the cells of the testes develop in the human embryo; in other words, they found the gene that determines gender (according to them). Researchers at the National Institute of Medical Research in London found a similar gene in mice. The discovery was said to have a potential use in animal breeding to control the sex of the offspring.[20] What about human reproduction?

For many years it has been possible to screen for sex during pregnancy, by looking at the chromosomes of foetal cells, or by looking at an ultrasound picture of a foetus in a woman's womb. It is now possible to screen IVF embryos for sex, just as it is possible to do so through amniocentesis, chorionic villus sampling or ultrasound scanning during pregnancy. Also, much research is being carried out on ways to separate girl-producing sperm from boy-producing sperm (biological sex is determined by the sperm). The medical rationale for doing so is that around 200 known genetic disorders are sex-linked; that is, they mostly affect one sex, either male or female, and rarely the other.

But sex selection methods are used in another social capacity: to eliminate female foetuses where daughters are unwanted or a liability, or to offer a choice of sex (girl or boy, you choose). In the USA, sex selection tests have been available through a biotechnology firm in Sausalito, California, which also offers them to clinics worldwide. There have been reports of sex selection in Australia, Britain, China, Korea and in India, the country where it has attracted the most publicity.

In India, clinics offering cheap services have mushroomed in rural areas where working-class people could avail themselves of it. Between 1979 and 1983, 78,000 female foetuses were aborted solely because of their sex. The women's movement in India criticises the three powers that allow sex determination to happen, the 'medical mafia', the ideology of 'son preference' and the population control doctrine.

In the *Illustrated Weekly of India*, Bharati Sadasivam explained:

Given the socio-cultural bias in this country it was inevitable that doctors were quick to realise its potential [of amniocentesis] as a method to determine the sex of an unborn child, and offered it to couples wanting to prevent the birth of 'unwanted' daughters. Sex determination clinics, with widely advertised services, have been set up in almost every city, town and suburb exclusively to treat 'patients' with one or more female children, while several gynaecologists offer the facility ... Specially significant is the place of sex selection in the country's population control programme. If sex determination tests can reduce the births of women who are 'breeders' of society, then, say family planning theorists, this could prove to be the most ideal and rational solution to the population explosion . . . [and] . . . To propose femicide for population control is to make the assumption that it is women who are responsible for the population boom and whose numbers should therefore be controlled by 'any desperate measures', and that the majority of them will persist in child-bearing until enough male children are born. Neither of these assumptions is borne out by statistics or experiences.

The problems, she adds, are lack of food, economic security, sanitary conditions and medical facilities. Sadasivam concluded from the personal testimonies of women who had undergone selective abortions of female foetuses that, 'To most of these women, amniocentesis is an unfamiliar and frightening word. To their husbands, it is one more means of authority and control.'[21]

Despite a national government ban since 1985 on the use of amniocentesis for sex determination in government hospitals, private hospitals and clinics continued to offer cheap sex screening services. As government authorities dragged their heels over the revelations, the Bombay-based women's group, the Forum Against Sex Determination and Sex Predetermination, carried out a campaign to increase public awareness and pressure the government to make the practice illegal. They cited the data collected by the government of Maharashtra State which showed that the number of clinics performing amniocentesis in and around Bombay city had gone up from ten in 1982 to 248 in 1986. These clinics were linked to 16 genetics laboratories which carry out the sex determination test; some of these laboratories received samples from towns outside Bombay. The business was profitable, and whereas some of the rural health centres did not have cold storage facilities for medicines, the means were found to keep the amniotic fluid (the fluid which surrounds the foetus in the woman's womb, and which is withdrawn by a needle for amniocentesis) in ice packs for the trip to the laboratories.

The Forum explained, 'They [doctors] tried to convince

us that they were doing a social service by catering to the psycho-social needs of people by making known the sex of the child and allowing them to make their own choice.' The Forum argues that sex selection for 'social' reasons – it's socially easier to have boys than girls in a world where girls are a burden – is something similar to the arguments made for other sorts of selection – it's socially easier not to have a disabled child in a world where disability is a burden.[22]

The state government eventually passed a law banning sex determination practices. However, clinics continue to operate, although not so openly, and at a higher fee. Advertising of the services in newspapers, which had been prevalent, has disappeared; but the tests are now conducted by special arrangement with doctors willing to refer couples to sex determination clinics.

Madhu Kishwar of the Delhi-based magazine *Manushi* estimates that sex selection practices – whatever they may be – will stop only when the social status of women is improved. When she investigated the correlation between social status and the sex ratio of females to males in the Indian population, she found:

> The twentieth century has seen a progressive decline in the sex ratio, resulting in a deficit of females in the population. The 1901 census showed 972 females per 1000 males. By 1981, there were only 935 females per 1000 males in the population. In this period, the overall status of females does not show signs of having improved. There is a growing gap between the literacy and employment rates of males and females. Female mortality rates continue to be higher than male rates, and several studies suggest that this is linked to neglect and devaluation of the health and lives of women and girls.[23]

When Kishwar compared the status of women in two states, Rajasthan and Kerala, she found a regional variation; low sex ratio and low status of women went together. In

Rajasthan, where women are considered an economic liability, where they have a low literacy rate and employment rate and where parents must pay a dowry at marriage, there is a lower sex ratio of females to males. By contrast, Kerala has the highest ratio of women to men, and here was found the highest literacy rate for women and a high employment rate.

The preference for sons, men's desires for sons, have been deeply respected throughout patriarchal histories. The social act of sex selection predates genetic screening; female infanticide, the killing of newborn baby girls, has been practised in many cultures at different times in history to eliminate unwanted female children. Genetic technology has not invented preference for male babies, but it has offered another outlet. As one headline in a British newspaper put it, 'Science gives Korea a new sex problem'. The problem was that sex determination was leading to a deficit of females in the population. A Seoul businessman quipped, 'In 20 years time, we'll have to import girls for them to marry.'[24] An entrepreneurial opportunity, perhaps?

Genetic engineering is not the fundamental cause of economic and social oppression, just as nuclear technology, for all its unique power to destroy the earth, is not the root cause of violence. But as transforming technologies they have both affected our perceptions of what is possible and 'normal'.

How can reproductive technology actually change women's lives for the better, enhancing women's social and economic status? No other transforming technology has done that – not the Green Revolution of mechanised high-intensive agriculture which was going to feed the world, nor nuclear technology, which was going to give cheap and non-polluting energy.

The genetic *priorities* of reproductive engineering are at loggerheads with social movements which conern themselves with empowering women and resisting discrimination. A social welfare approach requires a different set of medical, social and economic priorities. Unlike the new technologies, they will not create profits for pharmaceutical companies, they will not feed the burgeoning biotechnology industry and will not help win big scienfitic prizes. But they may give people more control over their lives.

The political protests over the creeping of biotechnologies into every aspect of people's lives began with the hope that it is early days yet. There is time to avert more crises. One need only look at the environmental and the food crises to see how drives for efficiency – be they motivated by profit seeking or by loyalty to an abstract idea – end up in practices that do harm.

It is here that the madness of the industrial development model comes full circle. The criticisms of it by women's groups all over the world are shared by many people. We have been called irrational. But what confidence can there be in a rationality that pursues the perfect baby alongside life-scorning technologies of progress, destroying the environment, contaminating the food, polluting the water and air, which are damaging the health and fertility of ourselves and our children.

9 Gene Maps

Protests have been strongest in West Germany with bitter memories of eugenics and a genetic vocabulary deeply tainted by Nazi abuse. Even the gene map – *Genkarte* – causes alarm as *Karte* is the name for the plastic bank and credit cards and there are fears of a 'gene card' that can be swiped through a terminal and checked as simply as a credit card.

Nigel Williams[1]

The project to create a global map of the human genes is rich in metaphor. It has been called a second French revolution, biology's moon shot, the holy grail of genetics, The Book of Life and The Book of Man. Its proper but less colourful name is the Human Genome Project or the Human Genome Initiative. It is taking place on many different levels, as the individual work of molecular biologists and as a major project supported by scientific and governmental organisations. It is said to be the single largest non-military science project today.

The genome project is 'Big Science', the term for scientific endeavours which require co-operative efforts and big money. Nuclear physics and astronomy, with their colossal accelerators and telescopes which are shared by scientists from all over the world, have been doing big science for years. Not content to 'think small', biology also has its own big project now in the genome initiative, though the genome project is seen more as a technical feat of engineering (mapping genes and decoding all the 'information' in human DNA) than of research science (exploring new vistas in the natural world). It is feasible because of gene manipulation capabilities and computer technology. Scientists backing it argue that if they knew what the entire human set of DNA was composed of, it would move science and particularly biological science to a new level of understanding. It would serve humankind.

A 'global map' of all the human genes. It is an interesting phrase. Maps are representations in outline of the surface features of lands, waterways, the planets, the stars in the skys. Or more generally, a map can be a representation or scheme of the disposition or state of anything. Every map says something about the view of the cartographer: about which things are represented, how they are represented and which things are not represented at all. In her prose autobiography *Zami*, poet Audre Lorde talks about Carriacou, the fondly remembered island home of her mother's aunt, a place she could not find in any atlas of the world or on any map until she was 26 years old. She found it in the *Atlas of the Encyclopaedia Britannica* 'which has always prided itself upon the accurate cartology of its colonies'.[2]

What exactly would a *map* of 'the human genome' be? What exactly can a human genome be?

The *genome* is the term geneticists use to mean the entire set of genes in an organism or a species. The human genome refers to the full set of genes 'necessary to make a human being'. Each human individual has a unique set of genes, except identical twins who inherit the same set. Your genome and my genome differ considerably, although each of us apparently has the same categories of genes within the chromosomes of our body cells. Hence scientists talk about 'the human genome', although they say there are ten million known genetic differences between any two individuals. When the question arose, 'Whose genes will be sequenced?', biologist Sydney Brenner, who heads a UK gene mapping team, said it would be the 'unknown genome' like the 'unknown soldier'. From the USA, not surprisingly, the suggestion arose to use the genome of a rich benefactor who might fund the project. The idea that the genetic role model would be a Rockefeller, Ford, Getty or Howard Hughes was amusing perhaps but not a serious option. At the Centre d'Etudes de Polymorphisme Humaine in Paris, the cells containing the genetic information of 60 large families from France, the US and Venezuela were collected, a representative sample which would presumably incorporate all known genetic differences, in anticipation of forming the 'reference material' for an international effort.

The human genome initiative is the project to (try to) make sense of all the human DNA. Technically speaking, it is the project to map and sequence (decipher the chemical structure) of all the human genes. The first step is to map the genes, that is to locate genes on the chromosomes; much of this is accomplished already, although there are many different sorts of gene maps of different degrees of precision. The second step is to find the sequence (chemical formula) of each DNA strand on the map. Only a few per cent of the DNA of chromosomes are genes (that is, only a few per cent of the DNA code for proteins); the rest of the DNA is involved in controlling the action of the genes, the packaging of DNA into chromosomes, and possibly other things. There is a big question mark as to what else is 'in' the DNA.

Although genetic engineering has made it possible to contemplate the grand scale of a human genome project, the practice of mapping genes predates it. In the early days of genetic science at the beginning of the twentieth century, sex-linked disorders were mapped to the X chromosome; this was the first sort of human gene map. It was accomplished through studies based on observations of traits in children. Later, molecular biology methods were put to use, allowing researchers to analyse specific regions along the chromosomes when looking for gene correlations to illnesses such as cystic fibrosis and Duchenne muscular dystrophy. These projects to 'find' genes associated with illnesses may be considered stepping stones to the big genome project.

It is a huge task, stupendously expensive, and time and resource consuming. An early estimate put it that to decipher the genetic alphabet of all the human DNA, with its estimated 50,000-100,000 genes, with a total of three billion (3000 million) base pairs, would take 30,000 person-years and US$2-3 billion. In 1986, Akiyaoshi Wada, professor of biophysics at Tokyo University, estimated that it would take 100 experts 250 years and US$1100 million; or, if done in a sequencing factory, it would take US$600 million and 30 years. In 1988 the US National Research Council recommended that $200 million should be provided per year for 15 years to find the sequence of the human genome.

Those estimates of cost fail to include the tremendous amounts spent on the planning stages and back-up expenditure. For instance, as Leslie Roberts states in 'Carving up the Human Genome' in *Science* the three-day Workshop on International Co-operation for the Human Genome Project in October 1988 in Valencia, Spain cost the Valencia government and industry more than US$1 million. At this meeting the ideal of international co-operation, of which there has been much self-congratulatory talk among proponents of the genome project, was shown to be far from reality. One scientist attending complained that by co-operation many geneticists mean competition. Scientists from developing countries have been especially concerned, indeed alarmed, at the prospect of the powerful nations having near-exclusive access to the 'universal genetic code'.

By 1989, time estimates were modified. If scientists and nations pooled their resources, the complete sequence might be assembled within 20 years.

Once the chemical formula is known, the work of

'sequencing the genome' is complete. But the sequence of chemical elements – called bases – says nothing about the function of that stretch of DNA. At this point the bigger task must be undertaken of 'interpreting' the sequence: figuring out what it means, finding the structure and function of proteins the genes code for, correlating biological data and human characteristics and behaviour. The genome project is at once very different from and not so different from nineteenth-century attempts to 'decipher the body', as with phrenology, the scientific study of the external conformation of the cranium which was supposed to shed light on the degree of development of mental faculties. The latter looked for correlations between the soul and the particular shape of the body. The proponents of the former, while not outwardly seeking knowledge of the soul through the structure and layout of all the genes, do share in the same tradition. At least, calling the project 'The Book of Man' or 'The Holy Grail', and making it out to be the key to the individual's 'disease state', implies as much. Both the differences and the similarities between the pseudoscience of the nineteenth century and the genetic science of the twentieth give cause for concern.

Everybody's Doing It

Three big genome efforts evolved at first and in tandem: the USA project 'Mapping and Sequencing the Human Genome'; Japan's 'Human Frontiers Science Program' (an international effort partly and eventually aimed at human gene sequencing); and the European Community's proposal initially called 'Predictive Medicine'. The USSR boasts its own high-priority genome project, and said it has established an 'institute of man' with the aim of combining knowledge of the biological and the social human being. Italy launched a genome project, while scientists in Britain, France and Germany have been working at the goal of gene mapping and sequencing. UNESCO (the United Nations Education, Scientific and Cultural Organisation) voiced interest in playing a major role in the international 'globalisation' of the genome project and offered to house the European office of HUGO, the international Human Genome Mapping Organisation launched in Geneva in 1989 to co-ordinate international research. (UNESCO's offer was refused.)

HUGO is a group of scientists from many different countries who are involved in genome sequencing projects, and who intend to play a role in advising governments on the scientific, ethical, social, legal and commercial implications of genetic research. US geneticist Victor McKusick of Johns Hopkins University, the proponent in the 1960s of the theory that all disease has a genetic component, became HUGO's president. With the formation of HUGO, scientists formalised their role in shaping the social uses of the genome project, and their power to influence social policy.

The establishment of HUGO was just about as far away from the initial responses to the Genome Project as could have been possible. Less than four years before it seemed another loony Californian idea. The first serious attempt to initiate it came from biologist Dr Robert Sinsheimer, then Chancellor of the University of California at Santa Cruz. The story is told that while Sinsheimer was looking for a big project to put his university on the scientific map, he dreamed up the Human Genome Institute. His idea failed to get the support of the National Institutes of Health, the major public funding agency for biomedical research in the USA.

The project was not very popular among many scientists. Miranda Robertson, an editor of the influential science journal *Nature*, wrote that pursuit of the genome *per se* was mindless, considering that:

If the skill and ingenuity of modern biology are already stretched to interpret sequences of known importance, such as those of the DMD [Duchenne muscular dystrophy] and CGD [chronic granulomatous disease] genes, what possible use could be made of more sequences?

Others considered it a waste of resources – money and support which would go to molecular biology, rather than other scientific disciplines. Two irked scientists commented:

> . . . it would hardly honour Darwin's memory, as the conceived sequencing project violates one of the most fundamental principles of modern biology: that species consist of variable populations of organisms . . . And what is to be done when the sequence is obtained?

In short, there can be no prototype genome. Also, the project was seen as more symbolic than useful.

> I do not believe that there is any strong scientific justification for knowing the sequence of the entire human genome. The motivation for doing it is frankly political, or 'science' political, more than it is that science is being held up by our lack of knowledge of every nucleotide (base pair) in the genome

criticised David Botstein, a molecular geneticist at the Massachusetts Institute of Technology. When apologists for the gene project made exuberant comparisons with physics and space science, Nobel Laureate David Baltimore took the analogy further. He said that the genome project would end in disaster, like 'the disaster in space science that we have today'.[3]

But someone was very interested. The genome project was adopted by the US Department of Energy (DoE), the former Atomic Energy Commission, the agency that built the first atomic bomb and now does civil and military research, including Star Wars weapons research, in three national laboratories.

The DoE have been interested in genetics, especially the mutating effects of radiation, ever since the first atomic bombs were dropped on Hiroshima and Nagasaki. They have sponsored studies of the genetic and heritable damage caused by nuclear radiation and other substances in the environment. This and their involvement in other aspects of genetic research, plus their vast technical resources, computing facilities, over-inflated budgets and their enthusiasm for Big Scientific Projects contributed to a favourable reception for the genome project. (The DoE had long been involved in related projects. The national laboratory at Los Alamos set up a gene bank, a huge library of genetic information available to researchers – a database of genetic sequences from any organism researchers care to study. Information for the data bank comes from reports sent in by researchers in the USA and at the European Molecular Biology Laboratories in Heidelberg.)

Some academic scientists baulked at the DoE's interest and involvement in the genome project. James Watson, the molecular biologist famous for his co-discovery of the chemical structure of DNA, complained that the DoE's

> principal interest has either been making weapons or supporting the high energy physics community. It's really an organisation of physicists and not of biologists. And physicists have immense egos and they think they can do biology better than biologists . . . having a physicist as a boss would rather scare me.[4]

He didn't have to worry about it, though, because in 1988 the Congressional Office of Technology Assessment suggested that the Genome Project become a joint undertaking of two federal agencies, the National Institutes of Health and the National Science Foundation. James Watson was eventually asked to co-ordinate the US effort and he is now the boss.

The DoE's interest turned a project which few wanted

into one that the scientific establishment wanted badly. I think that this happened for several reasons. Once the DoE became involved, others did not wish to be left 'behind' (as they saw it). Once advocates convincingly argued about its commercial and technological benefits, it became very attractive. Finally, all the disparate interests in finding genetic roots of the human condition were able to coalesce in the human genome project.

Why the Genome Project?

At this point, we may wonder about the motivations and objectives of a project which has been so controversial for so many reasons. The controversy has been many-layered because the project has multiple motivations and implications. The genome project is not only about scientific and medical interest in finding biological roots and genetic explanations for so many aspects of the human condition; it also has an attraction for its potential as human identi-fication technology – that an individual's genetic profile could identify that individual and could suggest many things about that individual; and it is also about techno-logical prowess.

The first argument in support of the genome project is always its potential medical benefits. It is necessary, pro-ponents argue, for the DNA-based medicine of the future; it will make it easier to locate genes, create medical diagnostic tests, more sorts of genetic screening and gene therapy. DNA-based medicine, they add, would need a new industry to drive it, one which could evolve from the genome project.

The human genome project turns the races for human genes associated with specific diseases into one big giant race to identify all the human genes. It assumes all of the problems of the genetic hypothesis, magnified. One of the scientists at the forefront of the work, Leroy Hood of the California Institute of Technology, projects:

I can readily evision the time in the future where every newborn will have a DNA profile done of many different genes that may potentially cause particular kinds of diseases, and that that information will be fed into a computer and the computer will say, here are the 40 things that you have to watch out about, that you are susceptible to, and even more important, here are 50 other things that we can do to counteract the effects of these possibilities.[5]

Interest in classifying all the human genes, however, is not by definition a medical interest. A medical argument can be made to justify any scientific project of biology. It is the acceptable face of science, especially of most molecular biological research on animals and humans. Genetic hereditary illness is the most acceptable motive for genetic research of any kind. But the genome project is considered first and foremost 'basic' or 'pure' research and as such it has other aims, for instance to study aspects of genetic evolution, embryo development and other questions in biology. By definition the project aims to explore the links between genetics and all human behaviour and character-istics (not just those associated with illness), and it is not uncommon to hear its practitioners speculate on how genome analysis will lead to the correlation of other characteristics to genes, such as musical or athletic ability or even shyness.

Two other motives, and certainly major reasons the genome project has gained the support of politicians, are that the long-term commercial spin-offs for drug companies are incalculable, and because it is deemed necessary to maintain a competitive edge in technology. New DNA probes and diagnostic equipment are expected to emerge from the work, and advocates of the project warn that any country not involved will end up largely dependent on imports from those technically advanced. The project is a justification for creating new computer data-processing and DNA technologies, and data banks to store the information. The Japanese electronics companies Seiko and Fuji have been involved in automating systems for DNA sequencing. Keen gene cartographer Leroy Hood explained that one of the features that attracted Congressional support in Washington

and in the end it may be the thing that really carries the day, is the idea that the Genome Initiative could serve as a stimulus for guaranteeing a supremacy in biotechnology for the next 20 or 30 years. That it would be the focus for spinning off entirely new kinds of industries. It would be the one arena in which America would seem to be certain of maintaining a supremacy.[6]

As a project that depends on advanced technological capabilities, it has another attraction. DNA pioneer James Watson enthused, 'Even if there weren't all the medical reasons, as a biologist I would say "why not do it?" It's too interesting not to.'[7] This is technological exuberance taken to its logical extreme. Yes, there is a joy in knowing, in discovering reality, in a rationality, in the words of Audre Lorde, which makes order out of chaos, which comes from observing patterns and rhythms. But technological exuberance is specific, what Arnold Pacey called the hidden purpose behind technological practice: the pursuit of existential joy, as exemplified in the words of Robert Oppenheimer, a physicist who took part in the creation of the first hydrogen bomb. Oppenheimer later recalled their goal as 'technically so sweet'.[8]

Since so many contemporary theories of life are located firmly in the genetic camp, it might be useful to place the genome project in the context of other biological classifications. Since the time of Aristotle, there has been a strand of natural science occupied with classifying all the living and non-living matter in the world. As the history of Western biological science is told, the living world may have seemed an unrelated mixture of plants and animals until 1753 when Karl von Linne published a classification of plants, followed in 1758 by a classification of animals. The Linnaean system, based on perceived similarities of basic structure (but not necessarily of function), is hierarchical. Some categories are included within or subordinate to other categories. Categories are ranked. For example, we, the species *Homo sapiens*, share the order *Primate* with a host of other species. Once an event or object is named, all other things can be perceived as either similar or dissimilar.

The weaknesses in the Linnaean system became apparent in modern evolutionary studies. Today, there is a plethora of different classification schemes based on different criteria of what constitutes a natural relationship among living things. Today, genetics is the discipline interested in the processes which result in the patterns observed by scientists studying the evolution of living things on earth. Taxonomists of before and today 'chop up' evolving lineages as best they can to inform the study of biology. And here is the rub: the chopping up of lineages follows from the premises of the science and the methods employed in contemporary studies of diversity – in short, how a classi-fication is made. Yet, such naming and categorising processes are conditions of creating scientific theories – for example, theories of evolution such as natural selection which states that differentiation in organisms occurs at the level of the gene over long periods of time.

By chopping up human DNA as best they can, by creating formal classifications of genes from hierarchies of simi-larities (which necessitate codifying differences too), by observing patterns in the DNA to order them for study, the second leg of the human genome project will be in keeping with a particular intellectual tradition in the natural sciences, and is just as full of presumptions and limitations (the most obvious limitation might be the technology available to chop up the DNA into segments of 'genes'). These aspects of classification systems old and new may say more about how biological theories and knowledge are created than about how the living world operates.

Biological Traps

If the classification of human genes is considered of such fundamental importance and interest to finding the origins and explanations of human characteristics and behaviour – whether in terms of health and illness, talents, intellect, aggression, or any of the other suggestions offered – will we not inhabit a world where an individual becomes the greatest threat to herself by virtue of her genome?

Questions of a new kind of eugenics and discrimination loom here perhaps more than anywhere else in the 'human genetics' debate. How can discrimination based on genetic labels be avoided if the thinking behind it brings out remarks like: 'Everybody is born genetically unequal'; or as James Watson offered: '. . . really to understand ourselves, we're going to really have to understand our DNA.'[9]

The dream has infinite scope. From the Centre for Cellular and Molecular Biology in Hyderabad, India, director Dr Pushpa Bhargava wrote, 'Once we know that [the sequence of human DNA], we would be in a position to know a great deal more about the individual or a group of individuals – including his "disease" status – than is possible today.' Continuing in the document 'Social Implications of Modern Biology', he considered that there is a genetic basis of both ethics and aesthetics. 'If altruism were not built in our genes, we would probably have been extinct!. . . We regard aesthetics today as an entirely personal matter. It may turn out that this really is not so. Indeed, there may be universal determinants of aesthetics which are built in one's genes.'[10]

The theory that there is a gene or genes 'for' altruism and other such human attributes has become popularly known under the rubric 'the selfish gene', the theory which puts forward that human behaviour is motivated by an (unconscious) drive to spread our genes as widely as possible. The man who popularised the concept in his book *The Selfish Gene*, ethologist Richard Dawkins, wrote that humans are 'survival machines . . . blindly programmed to preserve the selfish molecules known as genes'.[11] The selfish gene theory has been taken to extraordinary lengths, such as to defend surrogacy as a basic human drive, to explain why men rape women and to explain that people (supposedly) behave altruistically towards their relatives and aggressively towards non-relatives. In a variation of the theme, 'genes for bad motherhood' was suggested by one commentator to explain why baby seals are 'abandoned' on shores by their mothers.[12] These musings take a leaf from sociobiology, the discipline which ultimately posits that all human behaviour can be reduced to biological (meaning genetic) explanations. These examples and principles may seem extreme, but whether it is the 'less extreme' or the 'more extreme' hereditarian assumptions which are incorporated in the interpretation of gene analyses, it is legitimate to ask where they come from, how they are used, and how they can be used.

The genome project, which on the one hand is supposed to be about making a map out of the common biological determinants of the human species, on the other hand, establishes genetic differences between people. Some say that everyone must be vigilant that the new technology and knowledge is not abused. Some call for protective legislation to guard against discrimination on the grounds of genetic make-up, but it is difficult to imagine how this will be guaranteed if the built-in objective of the exercise is to pinpoint genetic difference. And that *is* the objective. Geneticist Eric Lander explained the project: 'The key to genetics is having genetic markers. Any sort of a genetic difference will do.'[13] Some day, will those identified as, say, musically and athletically talented be channelled in one direction, the 'untalented' in another, as is too often suggested? (Considering the track record of attempts to classify young children according to their talents and IQ, who can believe a genetic classifying scheme would be any less flawed, presumptuous and rife with bias?) What about people identified as having a predisposition to heart disease. Will insurance companies or employers claim cause for different treatment of these people? (There have been reports from the USA that some insurance companies are attempting this.) How far will the results of the genome project go to serve insurance companies, industrialists and the police in identifying individuals for whatever purposes they need – suggesting placing controls on people's lives, or at least willingness to adapt our lives to the projections someone makes from our individual DNA profiles?

Regarding the medical promises, the genome project not only scales up the hunt for 'disease-causing' genes. It is fuelling the shift to that kind of medical thinking. Margret Krannich and Annette Goerlich, formerly of the Women's Bureau of the Green European Link at the European Parliament, criticised the aims of the European genome project, subtitled 'Predictive Medicine', noting how the term 'multigene defects' is being used by the European Commission to label illnesses such as coronary artery diseases, cancer, stomach ulcers, rheumatoid arthritis, diabetes and serious psychoses. The formulation is revealing, they said, in that other factors – environmental, social, economic – disappear behind the classifications 'multigene defects', 'normal gene' and 'genetic defect'; it presupposes a standardisation of the genetic constitution of human beings and leads to individualising health risks. As governments are accepting these medical definitions, they worried, 'attempts to prevent an "overflow" of these technologies and their "misuse", without seeing that *any* use automatically implies misuse and that *any* attempt to draw "clear and legally established borders" is doomed to fail.'[14]

The director of Britain's Imperial Cancer Research Fund, Sir Walter Bodmer, lobbies for the genome project in terms of its promise to explain cancer and heart disease. One in three people in the country are likely to get cancer at some time in their lives. Genetic science is not primarily about hereditary conditions as we are used to thinking of them, he explains, but about the mechanisms of cancer and much more. In 1990, addressing the Parliamentary and Scientific Committee at the Houses of Parliament in London, Professor Bodmer promoted the public funding of a human genome project, arguing its ostensible medical benefits. He thus promoted a larger British commitment to gene mapping and sequencing apparently over and above the millions of pounds of public money the government had already committed. At the time, no eminent scientist was similarly lobbying for a revolution in public health through research on all the other factors which contribute to the illnesses. At the time, the National Health Service was fighting for its life, having been threatened by government cuts and changes in policy. At the time, social services and care for people who are ill and disabled were also faltering due to government policies. At the time, workers' compensation for disabilities caused by industrial accident had been shrinking. At the time, new evidence was emerging that children of fathers who work at the nuclear power plant at Sellafield have a higher chance of contracting leukaemia (a cancer) due to exposure to radiation. At the time, the UK's Ministry of Agriculture repeatedly advised that the chemical spray Alar, used on apples, was perfectly safe when it had been banned in the USA after it was directly linked with causing life-threatening cancerous tumours. DNA-based medicine suggests that it is possible and beneficial to look into controlling disease through the individual, not the substance.

Gene 'Fingerprints'

Full genetic profiles of individuals based on gene mapping

technology may seem some time off, but there is another sort of genetic profile that is already in use called DNA fingerprinting. It is used mostly in police work, or to prove familial relationship. It has been used to convict rapists and murderers; to reunite with their families some of the children who became victims of the former military dictatorship in Argentina; and to prove biological relationship which the British Home Office have demanded in certain cases of immigration. At first, DNA fingerprinting seemed failproof and advisable, but the circumstances of its use proved more complicated and troubling than was anticipated. For me these raised three issues: about confidence in the ability of science to create certainty from mystery; about how we live in relation to technology; and about authoritarianism and human identification technology.

State of the Art

DNA fingerprinting or genetic fingerprinting was a chance discovery of English geneticist Alec Jeffreys. In the early 1980s he was working at the private Lister Institute, studying regions along the DNA known as 'hypervariables'. He recognised that each person has a unique pattern to these regions, and that he could create 'gene probes' to identify them.

An individual's DNA is easily available for analysis, since most cells in the body carry a complete set of chromosomes, where the DNA is packaged. Any blood or tissue sample will do. The DNA is extracted, enzymes are used to chop the DNA into fragments. The fragments are separated out from each other according to (molecular) size. Gene probes are added which identify the hypervariable regions; the resulting profile appears as a series of dark bands on a gel or X-ray film surface.

Despite the connotation of the term 'DNA fingerprint', the profile obtained by Jeffreys' method is not of the total genetic material in a person's cells, and it is not a profile of 'genes' (pieces of DNA that code for proteins). The 'hypervariable' regions of DNA were discovered in 1980, and scientists do not know what their function is. Jeffreys' technique is only one type of gene profile. In principle, any part or the totality of a person's DNA can become a genetic profile. But that is a principle, which is why Jeffreys' discovery was so scientifically and commercially newsworthy.

The DNA fingerprint is said to be specific. The distribution of 'hypervariables' is unique for each person: they are inherited: half are shared with the mother, half with the father. The chances that two people will possess the same DNA fingerprint pattern is thirty billion to one according to Cellmark Diagnostics, a company formed to market the method. Jeffreys calculated that the chances are three times greater. Hence, the 'fingerprint' can identify an individual, or a genetic family relationship one or two generations apart (that is among parents, grandparents and children). Jeffreys stressed that there was no other sort of information hidden in the pattern, like race or hair colour.

The DNA test was patented by Jeffreys and the Lister Institute; within a few months the UK chemical giant ICI bought the world rights and created Cellmark Diagnostics with locations in Abingdon, England and in the state of Maryland, USA. ICI expects to earn millions from the method. Variations on the Jeffreys test from other researchers and new companies followed in quick succession. One commentator suggested that in future, DNA tests would be designed to tell the colour of eyes, race, and other identifying physical features.

DNA fingerprinting was immediately hailed as an example of how academic research can reap lucrative and practical rewards. At first the test was considered infallible. The uses it could be put to seemed endless, though mostly they were geared at policing of one form or another. 'Crime is a growth industry', as the narrator of a television documentary on DNA fingerprints explained.[15]

DNA in Forensics

In England and Wales, DNA fingerprinting was first used in a forensic capacity to identify an individual whose blood or semen samples were left at the scene of a crime. In January 1987 police in Leicester, England sought the killer of two teenage girls through mass screening of the men of two villages. They asked the 2000 local male residents to come forward voluntarily for elimination purposes. The murderer, Colin Pitchfork, was found in January 1988, but only after some confusion, since he got another man to give a blood sample in his place. He was given a life sentence.

The first time DNA fingerprinting was entered as evidence in a jury trial, on 5 October 1987 at Birmingham Crown Court in a rape case, problems with the technique began to emerge. A woman identified two men as the rapists. One was convicted and imprisoned for five years, and the second man was acquitted. The DNA fingerprint came from semen left on the woman, but it proved confusing to the case. The DNA fingerprinting method was not sensitive enough, said Home Office forensic scientist Dr David Werrett. He added that a forensic sample is not the perfect working sample of clinicians who can get fresh, large samples of blood or tissue; at this point, he added, the evidence from conventional fingerprint tests were just as conclusive as genetic 'fingerprints'. He said the tests needed to be improved.[16]

The use of DNA fingerprinting in this rape case demonstrated that the DNA association of a man's semen is not going to change the terms of rape trials. What is does provide is another type of identifying evidence. If the test is considered conclusive in identifying a man accused of rape, it provides evidence for the conviction of an accused rapist, as it did in a rape trial in Wales the following month.

The use of DNA fingerprinting to identify sex offenders and murderers, which one may find justifiable and welcome, might well be seen as a beginning of much wider use of it as a human identification method.

In 1987, the US National Crime Information Center suggested that computers carry genetic information on convicted criminals. Jeffreys was briefed by the FBI, the US Federal Bureau of Investigation; the FBI recommended that police nationwide use a standard method to facilitate exchange of information. In 1988 scientists at Cetus Corporation in California and at the University of California developed a much more sensitive technique which required only a single DNA molecule and could work on very old and degraded samples. When authorities in King County in the city of Seattle, Washington, proposed creating a data bank of DNA fingerprints of convicted sex offenders, the publication *geneWATCH* responded with an article by Donna Cremans, who summarised the uneasiness of many, including Dr Philip Bereano, who sat on the King County DNA fingerprinting Advisory Committee. The plan to type DNA from blood samples could be subjected to many other tests,

yielding far more information about a person than just identity. Who will have access to the data collected by the police . . . and will samples be destroyed after they are analysed? Will safeguards be in place to minimize sloppy or incorrect interpretations of test results? Since the probes employed in DNA typing are proprietary, will defendants attempting to clear themselves have to pay monopoly prices for DNA evidence? DNA fingerprinting could become part of a larger pattern of the expansion of intrusive testing of citizens by police and others in power.

Bereano said that he was worried about the creation of pariah categories of people who would be experimented on with the new procedures.[17]

Once DNA fingerprinting is adopted for some sorts of police work, it is likely to spread to other areas. In 1988, the British government voiced interest in using DNA fingerprinting on anyone considered a terrorist suspect in Northern Ireland, recommending that police be given the right to take swabs of mouth samples without the consent of the person suspected, with no indication of how the rights of innocent individuals would be safeguarded. Certainly, there has been a growth in the science and technology of political control which includes a growth in governments' demands for police technologies. Steve Wright of the Manchester Police Monitoring Unit in England analysed what he called the public order policing revolution and the police industrial complex for the magazine *Science for People*, a publication of the British Society for Social Responsibility in Science. Wright did not name DNA technology in particular, but listed among the 'hardware' of police technologies human identification technologies. Other examples of such 'hardware' were riot control weapons (such as rubber bullets), intruder detection equipment, computerised databanks, surveillance cameras, to name a few. These products are marketed by companies including the chemical company ICI, the car manufacturer Land Rover, the fireworks companies Schermuly and Brocks and a number of electronics companies such as Phillips, he said. He pointed to examples such as the paramilitarised state control which evolved in Northern Ireland as the major evidence of the technologies of repression in action.[18]

No one technique is enough to serve the programmes of mass control which Wright alluded to. But police acceptance of DNA fingerprinting, as a human identification technology which can be neatly adapted to computerisation and data banking, is occurring at a time in Britain and many other countries where police accountability to the general public has been dealt serious blows, when policing has been used to punish those who question the status quo. The implications are relevant to any assessment of the worth of DNA fingerprinting. But if the above issues raised about legal and state powers seem reparable in an ideal legal and policing system; if it seems that DNA fingerprints are necessary for responsible policing because they are super accurate and too useful to give up, consider the following incidents which undermined the scientific certainty of genetic identity technologies.

In the USA, a number of cases of questionable results arose as the use of DNA fingerprinting in police work increased. The first was a murder investigation in the Bronx area of New York City. A woman and her daughter had been killed. A local man, Jose Castro, was interrogated by the police acting on a tip. Lifecodes, a company carrying out the DNA tests, concluded that a bloodstain found on the watch of Jose Castro matched the DNA patterns of the woman. Lifecodes experts calculated that the frequency of the pattern in the Hispanic population was about one in 100,000. But fundamental problems with the assessment were uncovered – very technical details concerning the analysis of the blood sample – and what had seemed conclusive evidence could no longer be considered so. Eric Lander, the geneticist who analysed the data for the counsel for the defence, added that 'The general issues of interpretation are unique neither to the Castro case nor to Lifecodes.'[19]

Lander described three additional cases in which DNA fingerprinting would have brought about miscarriages of justice. One was of a woman accused of abandoning her newborn baby who was found dead, allegedly in the back seat of her car. Cellmark Diagnostics reported that its DNA analysis showed that she was the mother. She was not, as eventually became clear. A medical examiner determined that the baby was stillborn on 4 February 1988. The woman accused could never have given birth to that baby because she was in the early stages of pregnancy at the time, and gave birth the following October.

Lander pointed out several problems with DNA tests used

in forensics. Someone must determine if two unknown samples are identical. Reliability in recognising a match is technically demanding. The DNA fingerprinting results being used in criminal court cases in the USA are based on flimsy, inconclusive judgments. DNA fingerprinting also depends on inferences about the frequency with which matching patterns will be found by chance, which in turn rest on simplifying assumptions about population genetics. Contaminated probes are often used. 'It is my contention that DNA forensics sorely lacks adequate guidelines for the interpretation of results,' Eric Lander concluded.[20]

DNA fingerprinting, hailed as a powerful force of justice, can never be the final authority. These cases shook up those who had previously been so convinced of the accuracy and reliability of DNA fingerprinting. They also suggest that the new technology is not always necessarily going to serve women, although one of the other main targets of the new technology was paternity testing, hailed as a force of justice for women.

Policing Paternity?

Cellmark saw paternity testing as its most lucrative application and let it be known in women's magazines and the media that 'Women left holding the baby now have technology on their side.'[21] In the six months between June and December 1987, the Abingdon laboratory dealt with about 1800 cases to identify familial relationship, at £120.75 a sample. 30 per cent of these were paternity cases, and 60 per cent immigration cases.

What bothered me during all the media coverage of these two uses of gene profiling was that no one questioned the genetic concept of paternity and maternity that underlay the use of DNA fingerprinting to prove the parent-child relationship. This particular biological definition (a genetic one) is a confined definition of the family, of parenting and child caring. For one, what of children who may have been taken into a home, or adopted? Are they not real children? Or are they 'less' real because they are not biologically related by shared DNA?

While DNA fingerprinting might help identify some errant genetic fathers and make them pay child support, it is fraught with problems. If an individual woman wishes to make claims on the father through DNA tests, it looks as if she will have to foot the bill. It could also work against women. What repercussions would there be on women if the father wants to use DNA tests to see if in fact he is the genetic father with legal obligations towards individual children; or to claim custodial rights to children against the wishes of the mother. What if DNA testing revealed that the putative father is not the genetic father? On a policy level, it may put pressure on the man to assume a social role in the life of the woman which she does not want. Governments may embrace it to force errant fathers to take financial responsibility rather than for the state to provide support for children. This solution may not be a solution for the women and children; and on the level of social policy, it is regressive in that it expects women to be economically dependent on men by definition rather than enabling women to live their lives autonomously. The point is, DNA fingerprinting is presented as doing favours for women as a whole when in fact it rests on patriarchal assumptions which have in part caused the problems it is supposedly solving.

In the related use of DNA fingerprinting in immigration testing, a somewhat parallel situation arises of using scientific 'facts' to settle an issue which would better be solved by looking at why there is a problem in the first place. Britain is unique in demanding proof of family relationship for immigration purposes. Many immigrants are barred from bringing their children into the country until such a relationship is proven. The demand is mostly directed at people from Bangladesh and Pakistan. Although

immigration lawyers supported the use of DNA fingerprinting to help their clients, a spokeswoman for the Joint
Council of the Welfare of Immigrants pointed out that once
again, people from the Indian subcontinent are asked to go
through procedures they may not understand, and which
would not apply to immigrants coming from countries with
large white populations.

A confidential Home Office inquiry begun in 1986 into
DNA fingerprinting concluded that most people trying to
enter Britain are denied legitimate entry, and that of the 40
Bangladeshi-born and Pakistani-born men surveyed, 80 per
cent of the children they wished to bring into Britain
proved to be their genetic children. Hundreds of Asian
parents saw DNA fingerprinting as a chance to bring their
children into the country, but the Home Office and Law
Society were not eager to accept it. It was being used, but
mostly by people who could afford the fee. The Home
Office refused to pay the £100 per test even though this
cost is far less than that of their usual investigations which
include asking different relatives and neighbours in the
country of origin to draw up family trees. (By contrast, the
authorities jumped at the chance of using DNA fingerprinting to police Northern Ireland, a situation which
threatens individual safety and people's liberties.)[22]

Can DNA fingerprinting for immigration purposes foil
racist immigration policies? There remains the problem of
confining the definition of the mother-child-father relationship to a genetic one. In addition, proposals tabled in
1988 to change the handling of immigration cases raise
doubts about how far-reaching DNA fingerprinting will be.
One change alters the criteria for entry, requiring husbands
to prove that they can support and house their families
before they are allowed into the country.

On the other hand, there is a compelling success story
that is told about the use of biological markers, to locate
children of the disappeared, the political victims of the
military rule in Argentina between 1976 and 1983.

Identifying the Grandchildren

An estimated 9000 to 25,000 people disappeared during those seven years of military rule; people were kidnapped, tortured and murdered by the ruling juntas. Hundreds of their children disappeared:

> some murdered, some abducted, but most born in prison. The military followed a rule: to wait to kill a pregnant woman prisoner until after she gave birth – but the rule didn't apply to torture. The newborn children were then sold on the black market, given to adoption agencies, or taken as 'war booty' into military families

wrote Barbara Beckwith in the US magazine *Science for the People*.[23]

From the start of the atrocities, the relatives of the disappeared organised protests for the return of the survivors and the bodies of the dead. They organised as the human rights groups Mothers of the Plaza de Mayo, and Grandmothers of the Plaza de Mayo. The names come from the plaza in Buenos Aires where they have demonstrated every Thursday since 1977, carrying photographs of loved ones, wearing scarves on their heads with the names of their missing relatives and friends.

The Grandmothers have been tracking down the missing children with cunning effort, and finding some of them. But to restore them to their original families or to get visiting rights, they needed proof of the biological relationship. The Grandmothers approached US scientists for help, and in 1984 a team including forensic specialists and a geneticist travelled to Argentina to help identify the bodies of the disappeared and the children the Grandmothers found. The team could not trust the forensic specialists there, as most of them had worked for the military regimes. So they trained students in human rights organisations in forensic skills.

Many medical, scientific and forensic methods were used, in combination with eye-witness accounts and other kinds of evidence gathered by the Mothers and Grandmothers. DNA analysis was but one sort of 'genetic' marker to estimate biological ties; workers also looked at blood groups and the proteins associated with red and white blood cells which vary between individuals. This data, plus known distribution of blood types and other features in the population, were used to calculate an index of grand-parental relationship. The team estimated a probability of 99.9 per cent that shared genetic markers would be the result of family relationship, not random changes.

The team identified many of the disappeared, the causes of their deaths, the tortures they endured; they certified blood relationship between some of the children (but not all who were claimed) and grandparents; and their work led to the prosecution of many military officers, including a few high-ranking ones. Though the upheaval might have proved traumatic for the children involved, especially those who were living happily in adoptive homes for ten years, it seems that most of those children who have been restored to their original families thrive; they know the truth of their personal history.

A National Bank of Genetic Data was set up in 1987, to hold blood samples of relatives of disappeared children. The gene bank will also keep data on computer disk in an unnamed country outside Argentina – which seems all the more important due to a right-wing political backlash in Argentina, military pressure to forget or justify past actions, the granting of amnesties or time limits on court action against former crimes of the military and fears of a future military coup.

In the end, Beckwith noted, the team's forensic work set a precedent 'in pursuing those guilty of human rights violations, asking for outside expertise, accepting in-court genetic and forensic evidence, and pursuing the torturers and murderers of the disappeared', and it is an example of how a strong human rights network can counteract the military network of collaborators all over the world.

I want to make it clear that the technology of DNA fingerprinting is not the hero of the day, but that the strong human rights network – people who counteracted the forces of the military past and present – is. The success of this campaign does not erase the general and particular problems of biotechnology thus far. Next to the scale of what happened in Argentina, it would be feeble to suggest that it can redress the atrocities. In saying this, I would not wish to diminish the role of those with technical expertise in carrying it out. And neither do I think the use of molecular analysis in this case can be dismissed as irrelevant. It provided evidence of family relationships which could not be disputed by those who had more status and power in the society – the middle-class and rich adoptive parents and others who could have otherwise used their influence to maintain the lies about the children's origins (many innocently believed these lies were the truth). I believe this is an instance where a scientific technique made available through new gene technology was used well; another instance is where DNA fingerprinting may be used to identify perpetrators of violent crime. But we have to acknowledge that this particular DNA test would not exist if it were not for the present emphasis on genetics research and genetic engineering technology. But that emphasis historically evolved, not least because it serves industrial interests and the 'infinite growth' model of development, complete with questionable motivations, risk-laden products and processes and extremely troubling goals. But we have to ask again why we are left with only a few crumbs of real benefit, at a great price.

DNA fingerprinting has been put to good use, it may be said; but I do not think that proves a case for the 'revolution' in biotechnology and genetic engineering. When we are confronted with problems, there may be many paths along which to travel to find solutions. The complete scenario of

DNA fingerprinting holds a double meaning for me, one of hope and thankfulness that something positive has been accomplished, but one of warning, too.

For one, any emphasis on creating data banks of genetic information to secure a future where human rights are not violated – where people are not violated – is deeply flawed and unrealistic. If future governments undergo a repressive backlash, is the National Bank of Genetic Data out of the reach of forces of oppression? Data banks serve the cause of efficiency. It can be welcome when it means our own everyday dealings with official bureaucracies are managed better. But how much of this is short-term benefit in an otherwise bloated and dangerous system? State efficiency, as a theme, has run through twentieth-century analyses of fascism and state socialism, and their attempts at bureacratic population control.

How much of a technology is safe to embrace? How much can we control it? Commenting on police technologies, Steve Wright said that often they are meant to be hidden, appear safe when they are actually 'highly dangerous . . . Repression is most effective when its presence is ubiquitous yet invisible.' Or as another science analyst put it, 'To appear not to be about power is the most effective power of all.'[24]

You cannot talk human rights without talking about the wider picture of what is happening, or in this case, the effects of biotechnology elsewhere. Ecological principles are at work here, too. I hope that what I mean becomes clear in the next chapter, which concerns the militarisation of biotechnology, and a final analysis of gene ethics.

10 Gene Weapons

A new generation of chemical and biological weapons is crawling out of the test tubes of genetic engineers. It's like a Second Coming of the atomic bomb – and it could revolutionize war in ways even the generals don't like –
John Hubner[1]

Militarism is a form of colonisation which takes away from our lives.
Penote Ben Michel, at a conference in 1987 on militarism in Labrador/Nitassinan.[2]

Genetic engineering did not start out as a weapons technology, but it has ended up there, despite the fact that biological warfare is outlawed by international treaty. No one knows the full extent of military involvement, but an article in *Bulletin of the Atomic Scientists* warned in 1983, 'A misuse of bioengineering technology for military purposes seems probable unless a re-examination of and a stronger commitment to biological weapons disarmament is forthcoming.'[3]

It has been said that any technology can serve the military in a militaristic society. I take this point to an extent. But I do not think it goes far enough in explaining the relationship between biotechnology and the military, just as to say it would not adequately explain the civil uses of nuclear technology, such as in nuclear power plants and food irradiation. I do not want to 'blame' the technology, however, nor would I say that the ultimate stimulus towards biotechnology is in its biological warfare potential. On the other hand, the militarisation of biology and biotechnology is not simply a matter of ransacking a cupboard for exploitable technologies, and by sheer coincidence finding something in biological engineering. The impact of science on military affairs is as important as the impact of the military on scientific affairs. Analyst John Hubner concluded, 'At this point it appears the military will be as integral to the growth of biotechnology as it was to the development of the aircraft industry.'[4]

Perhaps a useful way to think about it is in terms of the recognition of the strong relations among science, technology, industry, government and the military, as articulated by Dwight D Eisenhower, former army general and US president from 1953 to 1961, in his parting speech to the citizens of the United States. He warned of the danger to democracy posed by the 'military-industrial complex' and the 'scientific-technological elite'. Later, Solly Zuckerman, a former adviser on defence to British governments, confirmed, 'The closest possible connection exists, and will always exist, between science and military affairs.' Considering the technology connection, Arnold Pacey wrote in *The Culture of Technology*, 'Scientists' pressure groups, defence bureaucracies and large-scale industry wield power almost beyond political control.'[5]

Preparing for war is a prosperous business for certain vested interests, and justifies commitment to particular kinds of technology. Each technology has its own historical development. Genetic engineering was invented and developed by academic researchers working in academic laboratories often in partnership with industry, unlike nuclear technology which is a direct descendant of the military project to build the first hydrogen bomb. But the relationship between genetic engineering and the military began earlier; and military interest underlines genetic engineering more heavily than may be apparent at first sight.

For one, the seeds of the biotechnology industry were arguably sown during the First World War when the British government sought an alternative source of the chemical acetone, which was necessary for the manufacture of explosives. The successful alternative process exploited the properties of a bacterium which produced the raw chemical. That process became the basis of industrial fermentation and thus the birth of the biotechnology industry, the latest home of gene manipulation technology.

Linda Bullard, formerly of the Gen-Ethic Network in Berlin, placed it thus:

Genetic engineering itself is in a very real sense the direct descendant of the Bomb. When we entered the Atomic Age with the bombing of Hiroshima and Nagasaki in 1945 scientists and some government officials realized that radiation could have harmful effects on human gene mutation. This sparked a renewed interest in genetics research, and the Atomic Energy Commission began to pour funds into this field. One of their grants went to James Watson and Francis Crick, who, in 1953, unraveled the molecular structure of DNA, the key to genetic engineering . . .[and]. . . the military also seems to have developed an interest in the new reproduction technologies. In 1984 NATO organized a week-long seminar in *in vitro* fertilization led by the renowned father of the first French test-tube baby, Jacques Testart. Perhaps on of the reasons is that IVF provides a source of surplus embryos which could be valuable for BW [biogical warfare] research. We do know that during the 1970s foetuses imported from South Korea were used for this purpose during a time when the US BW program had been cut to almost nothing. But even then there was sufficient activity to utilize 12,000 pairs of kidneys from three-month, old foetuses produced by Caesarean section. During the same period the number of abortions in South Korea rose to three times the number of births.[6]

In the last decade, there has been a revived military interest in biological research, in particular in capabilities seen to be offered by gene and cell manipulation. Let us consider these developments: what gene weapons are or may be; and the uneasy alliance among the military, industry and science worlds.

Biowarfare in History

The use of biological agents for warfare is as old as civilisation, which is to say, in patriarchal documentation of it. Historians date instances of the deliberate spreading of disease back at least 2000 years in Western history, citing cases in which ancient Greek and Roman warriors poisoned the drinking water of their enemies with decomposing corpses. In fourteenth-century Europe, the plague was used against the population of cities by throwing the corpses of its victims over the city walls. The smallpox virus was used against Native North Americans in the eighteenth century by the British under Jeffrey Amherst, when one Captain Ecuyer intended to cause an epidemic among Native people when he gave two contaminated blankets and a handkerchief to two tribal chiefs.

In the twentieth century chemical and biological weapons – like 'conventional' or nuclear ones – are weapons of the technologically literate and dominant. Leonhard Wolfe defines biological warfare as

the intentional use or manufacture by culture or cloning of disease-producing viruses, bacteria, fungi, insects, or toxins produced by these organisms, for the purpose of causing disease or death of man, animals and plants.[7]

Sometimes a distinction is made between a biological weapon and a toxin weapon. The biological agent is a living organism, and one which can be discharged in spore or other dormant form. A toxin is a poisonous substance of biological origin. For example, botulin is the potent toxin produced by the bacterium *Clostridium botulinum* which leads to the severe and often fatal poisoning known as botulism. Hence, a toxin weapon can be understood as a chemical substance of natural origins such as a bacterial poison or snake venom, and the boundary between chemical weapon and biological weapon is not always clear. Are toxins a special category of chemical weapon, or are they biological because of how they are made?

Chemical and biological weapons may be directed against people, animals, plants or the ecosystem which supports the lives of people, or they can be directed against equipment on which military or civilian personnel depend (for example, bacteria which chew up oil can wreck machinery). Pathogenic or disease-causing micro-organisms studied as potential 'anti-personnel' agents include the viruses for yellow fever, Rift Valley fever, smallpox, influenza and encephalitis; the bacteria associated with plague, anthrax, cholera, dysentery, and typhoid; the rickettsiae which cause epidemic typhus and Rocky Mountain spotted fever, to name a few.

The first time chemical weapons were used on a large scale was in the First World War, when chlorine and mustard gas weapons were developed; these cause harm when inhaled. Germany used them first, and Britain, France and the United States developed their own versions. In this war, one million people were killed from chemical agents, whose development required a chemical industry. In the 1930s, nerve gas, which harms when absorbed through the skin, was developed in Germany. Around the same time period, the US, UK and Japan concentrated on developing biological weapons. The British experimented with anthrax by infecting the whole of Gruinard Island, which lies off the northwest coast of Scotland. It remained uninhabitable until recently, when the government carried out a decontamination programme. During the Second World War the Japanese military directly experimented on human subjects, using biological agents during attacks on China, and on thousands of prisoners of war, mostly Russians and Chinese and also US and European prisoners. At the end of the war, the US military offered Japanese scientists immunity from prosecution in exchange for information about those experiments. The decade after the Second World War saw the invention of more chemical agents. In the UK, the Chemical Defence Experimental Station at Porton Down picked up the work previously done by the chemical company ICI. At Porton Down new types of nerve gas, tear gas and defoliants were created. The latter were subsequently used in the 1960s in Malaysia to destroy plant life. Agent Orange, a mixture of two chemical herbicides, was the most prominent anti-plant agent used by the United States in the Second Indo-China War in Vietnam. Defoliants or anti-plant agents were employed to destroy food crops and forests, to expose guerrillas and drive the rural civil population out of their habitats and into US-controlled urban areas. The large-scale use of anti-plant agents destroyed much of the upland and coastal forest ecosystems of South Vietnam.[8]

Horror at the destruction and death wrought by US military technology in Vietnam fuelled anti-war protests which made chemical and biological warfare a political issue that could not be ignored by national governments and the United Nations. It led to a unique disarmament agreement, the Biological and Toxin Weapons Convention 1972, the only disarmament treaty which bans possession of a whole class of weapons of mass destruction. It strengthened the Geneva Protocol of 1925, which prohibited the first use of chemical and biological weapons, but not producing, stockpiling, or experimenting with them. The Biological and Toxin Weapons Convention went further, banning research on and development of biological weapons and their stock-piling. It allows some research on biological agents for 'defence' purposes, which many analysts interpret as inconsistent with the general aims of the treaty. Offensive and defensive studies on pathogenic organisms are indistinguishable.

By the time the treaty was completed and signed by politicians such as US President Richard Nixon in 1975, biological weapons were known to be particularly inefficient and unpredictable, and hence less attractive to the military. Unlike any other weapon, live organisms reproduce and mutate. The users of the weapons could be as vulnerable as the intended recipients. Live organisms, spores or airborne poisons may move around to places they were not meant to go. 'Even the very concept of defending against such agents is misleading' is the general conclusion, considering the unpredictability of the spread of disease and the large range of biological agents which could be used. The unique unpredictability of biological weapons remains today. Genetic engineering does not change this characteristic of biological agents.

Bioweapons Today

None the less the military is definitely interested, perhaps because, from a military view, biological weapons have an advantage over nuclear weapons in that they do not have to be stockpiled. A small store of a culture of organisms in the laboratory will do; large quantities of organisms can be grown quickly, then put in a delivery system which could be something as simple as an aerosol sprayer. Biological weapons are called the 'poor man's nuclear bomb'. Creating lethal organisms (bacteria, viruses, etc.) and biologicals is apparently less capital intensive than creating nuclear weapons. US Deputy Assistant Secretary of Defense for Negotiations Policy, Douglas Feith, referred to these weapons as 'easy, fast, clean, and versatile', saying, 'It is now clear, using the newest technology, that BW can be highly militarily significant. We have changed our opinion about the military utility of BW.'[9]

Steven Rose, biologist and co-author of *No Fire, No Thunder: The Threat of Chemical and Biological Weapons*, surmised that the renewed military interest in biological warfare and genetic engineering in particular was fuelled partly by biologists who were over-enthusiastic about the potentials of biotechnology.[10] He has said that biotechnology immeasurably strengthened the university-industrial connection, bringing greater secrecy through the lure of patents. Inevitably, Rose suggested, all this activity was going to interest the military. What were they missing out on? they must have wondered. It was destined to attract scientific entrepreneurs unable to resist the combination of power and grant money that the military work offers. (I would add that sometimes scientists find themselves depending on military funding when other sources dry up at certain critical moments in budgeting. Some researchers have left projects when military links became evident.)

The following are some examples of what genetic engineering offers in weaponry, in principle at least.

- The variety of pathogenic organisms that could be constructed using gene manipulation technology is in principle limitless. For instance, genes which code for a toxin (poison) may be grafted into a commonly occurring bacterium such as *E. coli*, which inhabits the human gut; when infected, our bodies might not be able to recognise them as dangerous.
- Bacteria or viruses with altered genetic make-ups could be more powerful, more virulent, and resistant to antibiotics.
- Designer diseases without known vaccines could be created.
- Specific antidotes or vaccines could be produced to

pathogen and the vaccine would be unavailable to the enemy.

- Genetic engineering of bacteria could be used to manufacture large quantities of the most lethal toxins, or toxins which hitherto could not be produced on a large scale; toxins could be altered molecularly to enhance their destructive effect. For example, the gene which codes for the deadly toxin anthrax could be isolated from the bacterium in which it occurs, cloned and transferred to E. coli; or protein engineering could be used to study or alter botulism toxin; or the genes which code for toxins from cholera or diphtheria bacteria could be sequenced.

- Finally, so-called ethnic weapons – made to selectively incapacitate or kill certain racial groups – may be possible if researchers could identify genes specific to certain racial groups. However, there is no solid evidence to suggest that ethnic weapons have been developed or researched. Where the 'ethnic gene weapon' idea originated I do not know, but I think the prospect of making this sort of weapon specific to a certain group is unlikely; genes are not static, genes do not act in isolation, and genes may not even be relevant here – that is, if a bioweapon is a risk to some particular group of people, it may be a risk for all. I mention the idea because it often is raised, and does illustrate the power of gene-think.

Which, if any, of these military prospects are being carried out? It is difficult to say, since so much military research is classified top secret, but the United States' Freedom of Information Act has allowed persistent researchers to uncover some of that activity. (The UK has an Official Secrets Act which works to the opposite end.)

In the early 1980s, US Army scientists were working on isolating the genes of anthrax, sleeping sickness and typhus, at Fort Detrick and the Walter Reed Army Institute for Research; the Department of Defense was contracting out to universities for similar work on different diseases.

The Council for Responsible Genetics (formerly The Committee for Responsible Genetics) compiled a list in 1985 of 60 out of a total of 130 contracts which the Pentagon placed directly relating to genetic engineering. These are in addition to projects which are said to assist defence, for example creating a vaccine for Rift Valley fever. Here the lack of distinction between offensive and defensive research on biological agents becomes clear. Knowledge gained for defence purposes serves offence purposes as well. 'In order to create a vaccine, you have to be very familiar with a toxic organism. That is knowledege you would want to have if your objective were to build offensive weapons,' Robert Sinsheimer has said.[11]

By 1987, wrote Nachama Wilker and Nancy Connell in *geneWATCH*, an increasing number of biologists and biochemists were involved in research in the US Biological Defense Program.

> Overall funding for this program has increased by nore than 400 per cent during the current administration . . . Over this same period, federal support for research into civilian-oriented and basic scientific questions has sharply declined . . . Most of the research funded through this program focuses on the study of pathogenic organisms – viruses, bacteria and protozoa – and on the development of vaccines against them. Some of these diseases are exotic and deadly, while others are endemic to regions of strategic interest to the current administration's foreign policy, such as Africa, South and Central America, and Southeast Asia.

At the time, the programme had contracts with about 300 industrial, university and research centre laboratories in the United States. The Foundation on Economic Trends' president, Jeremy Rifkin, later reported that some US military contracts were finding their way to other countries, like Britain, where they could be kept secret.[12]

In 1989, the US Department of Defense named bio-technology in its list of critical technologies. Genetic Biology is one of three divisions of the US Department of the Navy's Research Unit; they established a recombinant DNA programme in the 1980s.

Charles Pillar, co-author of *Gene Wars*, warned in 1983 of actual military research undertakings in these areas in the United States.

> . . . under the guise of 'defensive' research . . . the Department of Defense has sponsored a broad program of studies, applying the latest techniques of genetic engineering in ways that could give the U.S. military the capability to wage biological war. The results of that research could send the arms race off on another dangerous spiral.

He found no evidence that gene splicing and cell fusion were being used to create designer organisms or to enhance the virulence of existing organisms and toxins. Rather, he said, the main thrust of research at that time was to develop vaccines to protect US and allied troops and populations against biological agents that might be used by someone else. Gene splicing is used to create the new vaccines, while cell fusion technology can purify the vaccine faster and better than previous methods.

> The next step is mass inoculation. Here a problem arises . . . mass vaccination for rare or exotic diseases would touch off a public furore, so the Army apparently has another idea. It is studying the effectiveness of aerosol immunization, in which the vaccine, against such potential B.W. agents as anthrax and tularemia, is inhaled rather than swallowed or taken by injection.

He estimated that this mode of delivery is more dangerous

and more likely to spread disease, but it can be used secretly, administered to an entire civilian population covertly. Lest the reader think the idea far-fetched, Piller quotes from a 1963 article in the army's publication *Military Medicine*:

> 'a plan for large-scale immunoprophylaxis of the civilian population should be prepared. This would include standby legislation for compulsory immunization if required.' A separate article cites aerosol vaccination as a means to accomplish that goal.[13]

Pillar's exposé overshadows a remark made by Colonel David L Huxsoll, general commander of the US Army Medical Research Institute of Infectious Diseases. Commenting on the research done at the military laboratories at Fort Detrick, Maryland, he said, 'What we do here is no different from what is done at Merck and Company [chemical corporation], or NIH [National Institutes of Health] or CDC [Center for Disease Control].'[14]

For the UK there is no published list of Ministry of Defence (MoD) contracts, but in 1987 there were about 750 contracts with universities and polytechnics covering all of its interests, and of these 65 were tracked down as covering research in chemical and biological warfare.[15]

Arthur Westing, formerly of the International Peace Research Institute in Oslo, Norway, believes that cultural constraints and revulsions will minimise the threat of biological warfare in our future. Well versed in the long history of biological and chemical weapons, he cites the 'virtual non-use of micro-organisms as weapons' as 'the primary example of such restraint ... Only four nations are known to have possessed micro-organisms as weapons in the past: France, Japan, UK and the USA. None admits to possessing them today, nor is there evidence to suggest otherwise.'[16]

The Biological and Toxin Weapons Convention was renewed in 1986, incorporating greater controls, including improved verification of activity, expanding definition of toxins, making the treaty part of the civil legal code of each country who signed it, declassifying research. Bullard affirms that the treaty was significantly strengthened but 'if the rulers of the world are this concerned about the proliferation of biological weapons, the situation must be worse than we imagined. In any case, treaties are only as reliable as the men who use them, and that gives us little cause for complacency.' After attending the second review conference of the Convention, US biologist Barbara Hatch Rosenberg wrote, 'It was apparent to this observer that more participation by the scientific community would make a difference; yet scientists hold back.' Similarly, Jonathan King said that 'the resistance to militarization [of biomedical research] has been slow in developing, due to efforts of certain sectors of the research community and of the biotechnology industry to deny the existence of special hazards associated with biotechnology.'[17] This interesting comment is well worth considering further, since safety is one of the key controversies over biotechnology in general.

Oversights

Whether biological weapons are used or not, the question of safety remains and this is the place to ask what distinctions can actually be made between military and civil biotechnology. A number of cases in the US over the second half of the twentieth century bode ill. A report from a US Senate committee accused the govenment's biological defence programme of 'lax safety enforcement' in its supervision of the 100 or so contracts in industry and academia. The Foundation on Economic Trends in Washington DC brought two suits against the Department of Defense on behalf of the public, one on the grounds that the entire biological defence programme posed a risk to public health and the environment. They charged that by definition the pursuit of bioweapons is unsafe because of risk of accidental release, theft, use of such biological agents by terrorist groups and anyone else who does not feel bound by treaties. They cited a case in which a large quantity of dangerous virus went missing from an army biowarfare laboratory at Fort Detrick in 1981, and in another suit charged the army and DoD with failing to investigate the disappearance of 2352 millitres of chikungunya virus (which could infect the population of the world many times over).[18]

In the 1950s and 1960s, the US Army carried out germ warfare tests in populated areas of the country, exposing millions of people to bacterial and chemical agents, and without informing them or monitoring the health effects of the tests. The purpose was to make observations of the army's capabilities to detect bacteria and their dissemination patterns. The exposed citizens were not considered experimental subjects of the experiments. Leonard Cole, author of *Clouds of Secrecy: The Army's Germ Warfare Tests over Populated Areas*, reported that 239 populated areas throughout the country were sprayed, including San Francisco, St Louis, Key West, Washington DC and the New York city subway system. A chemical, zinc cadmium sulphide, was used to approximate the activity of bacterial agents. The bacteria used were supposedly benign, but most of them, 'while less harmful than actual biological agents, were all known to cause disease'.[19]

In 1986, the US Army planned 'contained aerosol' studies of extremely hazardous viruses and other biological materials, as well as aerial spraying with less toxic elements at a military installation in Utah. Cole made a connection between these activities and open-air testing of genetically modified organisms for industrial and academic reasons.

Approval of open air tests with genetically modified organisms can only encourage the notion that biological warfare vulnerability testing is safe and, in the army's view, no less important than any other program. Second, the quest for an ideal simulant for outdoor testing may encourage the army to turn to recombinant DNA technology.

Barbara Hatch Rosenberg further pointed out that

No matter how closely such facilities are guarded, it is hard to justify the experimental creation of new and highly dangerous pathogens, even if they are useless as weapons, under the rubric of 'protective purposes'. Their very existence would be an insult to life on earth. Steps to prevent this from happening need to take place as soon as possible.[20]

The research itself is dangerous. Closed facilities harbouring lethal germs are dangerous. Open-air testing, even of apparently non-pathogenic organisms, is dangerous.

One must ask why civilian research incorporating the release of genetically modified micro-organisms is not considered an unsafe experiment on human populations. In the *Bulletin of the Atomic Scientists*, science historian Susan Wright and biophysicist Robert Sinsheimer wrote, 'Deliberate construction of harmful biological agents has generally been acknowledged as the most extreme biohazard associated with recombinant DNA technology.' *What of the seemingly less extreme ones?* They likened the situation in 1983 to the field of nuclear weapons in 1940 and the failure of efforts to prevent an arms race and its perilous consequences, saying, 'The use of the accumulated knowledge of biology for the construction of deadly pestilence is an ultimate perversion.' *What about the penultimate perversion, and the one before that, and the one before that?* They make a distinction between applying the new biology for peaceful and for military purposes. They suggest all governments agree to 'renounce research into agents of pestilence except for peaceful purposes, with all appropriate research to be conducted in internationally registered laboratories, open to scientists of all nations . . . One window upon Apocalypse is more than enough.'[21]

I would not wish to undermine the work and commitment of these and other scientists and historians or anyone seeking to uncover and resist the activities of those interested in using biological science and technology for military purposes; they have played a major role in alerting us to the dangers of military technologies (and not just biological agents). Nor would I deny that military research expressly aimed at creating lethal organisms and poisons is far and away the most insulting and life-scorning side of biotechnology. Certainly I would support disarmament efforts and the need to openly challenge direct military involvement before military dependencies build up, and the need for true openness. But the problem I find with the above assessment is in the definitive distinction it makes between the safety of military and civil research, or the category 'research for peaceful purposes'. A biohazard, whether it comes out of a military laboratory or a civilian laboratory, is a biohazard. There is more than one window upon Apocalypse already.

Jonathan King said:

> The development of a biological arms race would pose a grave danger not only to our national security but to the security of the entire species, just as the unleashing of nuclear weapons now endangers all of human civilization.[22]

Yes, but life on earth faces extinction even if nuclear bombs are never again dropped in war, as Rosalie Bertell demonstrated in her study *No Immediate Danger: Prognosis for a Radioactive Earth*. Bertell, a nun, scientist and peace activist, argues that the biological effects from so-called low-level contamination of penetrating ionising radiation in the atmosphere is bad enough. The increased mining of uranium and the nuclear energy power installations which emit radioactivity are initiating a death crisis of the human species. Legal limits on radioactivity, Bertell has calculated, are not safe limits. Indeed no level is safe, she argued, Bertell has chronicled the millions of casualties already, among them lesser-known health risks, harmful effects such as enzymes disorders in children and infertility when background radiation in the environment is high (called low by many authorities who are quick to point out that radioactivity is also a naturally occurring substance always present in our environments – though from that fact it hardly follows, as they intend, that adding more to the environment is acceptable). It has been in the interests of governments and industries not to monitor the true effects of such radiation.

Ironically, increased radiation contamination in the air, water and soil is considered an environmental stress – a threat – to the human gene pool. Damaging gene mutations, a deleterious effect of penetrating ionising radiation, is affecting the human population.

In her study, Bertell went on to look at the present conditions in the country of Peru, where development along the plans of the international financiers known as the International Monetary Fund came to mean for many death by malnutrition and starvation; for example people living in the slums buy water from private vendors while companies like Coca-Cola pay low prices for water, which is in short supply. She laments that the Peruvian government bought a 'false dream of remnant survival', a nuclear reactor, purchased from the US for over US$80 million.

There is more than one window upon Apocalypse already.

The world's worst industrial accident occurred between 2 and 3 December 1984, in Bhopal, India, when the lethal gas methylisocyanate escaped after an explosion at the Union Carbide plant. The death toll was first estimated at 3150 people, with 300,000–500,000 people injured. It is likely that these figures are low, and unofficial estimates put the death toll at 30,000 people. Six years later most of the people in the surrounding slums, which were the worst affected areas, remained ill or disabled; medical reports indicate increasing sickness and mutagenic effects, that is, harmful genetic mutation.

Although Union Carbide defends its response to the disaster and says its attempts at prompt action were held up through the neglect of the Indian government, the company was primarily interested in saving its image. The Indian government seemingly identified with the company because of its industrial development interests. Union Carbide made deals with the government to minimise their liability to those who were injured and the families of those who died. Then, in November 1990, Union Carbide was due to participate in the conference 'Corporate Responsibility and Environmental Excellence' in London, boasting its ability to teach other companies how 'to take the heat out of the environment debate about business as a polluter'. (They pulled out shortly before the meeting, which may or may not have been due to the protests directed against a number of the companies participating on the grounds that they have a record of putting profit before 'environmental excellence'.)[23] Union Carbide arguably is one of the leaders in corporate responsibility concerning environmental safety. That is the whole point. If such a disaster as Bhopal could happen in a Union Carbide plant, their high level of excellence in these matters is not high enough, and is cause for concern in the industry as a whole.

The inability to honestly appraise the risks and the harm brought about by nuclear technology, the chemical industry, the maldevelopment model with its 'wilful bleeding of people' (Bertell's phrase) – these are indicated in the advocacy of biotechnology. There comes a point at which to focus separately on the hideous effects of biological agents of warfare as a unique and contained category is debatable. The Apocalypse of which Wright and Sinsheimer warn can happen without a biological arms race. One can appreciate the difference in speed, probability and perhaps magnitude of a military-produced disaster. None the less, species annihilation is species annihilation. The chances of a single genetically engineered organism causing total destruction may be slight, as Cary Fowler, Eva Lachkovics, Pat Mooney and Hope Shand expressed in the 1988 issue of *development dialogue* entitled 'The Laws of Life: Another Development and the New Biotechnologies', but, 'There will be accidents. There may not be a "genetic Chernobyl" . . . but we might not know if there were.'[24] There comes a point at which you can understand an intention to harm no matter how deeply buried is that intention.

> Rachel Carson once wrote that agricultural chemicals were a stick hurled against the fabric of life. Biotechnology may give us life hurled at the fabric of life. Unlike the chain reaction in a nuclear power plant – where the elements are contained and controlled – the release of genetically-manipulated organisms could launch a chain reaction which we can neither understand nor control.

That is one fact of life. A second is: 'Any new technology introduced into a society which is not fundamentally just will exacerbate the disparities between rich and poor.'[25]

How peace-loving and life-enhancing can biotechnology be if these are its consequences, no matter whether you see 'life hurled at the fabric of life' as one side of a two-sided situation in which there are both good uses and bad applications of biotechnology; or, as I do, as a Möbius strip where there is only one side but that has various twists and turns.

Epilogue   Life After Genes

Tomorrow has no fixed address.[1]

I am sharing a table with a policy advisor to the United Nation's Food and Agricultural Organisation, Dr B S Raghaven, formerly of the Indian Administrative Service, who is visiting England. He talks about the Trieste-based international peace and food organisation of which he is a member. Eloquently, he talks about respect for mother nature and the earth. On the table, he draws imaginary concentric circles: food for all, then peace for all, then health for all, better productivity. I ask about the two big new international biotechnology institutes based in Trieste and Delhi. The conversation turns to biotechnologically produced products. Dr Raghaven suggests that biogenetic alternatives may be a viable answer to many problems. I say that I think they pose the same problems, and talk about the bio-pesticide business, the controversial packaging of genetically engineered seeds with chemical pesticides, the unknown risks. Dr Raghaven says we don't know yet. You have to test the products, carefully and slowly. This is the scientific spirit. He doesn't agree with the marriage of chemical pesticides and biogenetics. He would not have such a thing happen. (He must realise, as I learned, that herbicide tolerance strategies, such as herbicide tolerant seeds, encourage more use of toxic products in places where they may not have been used in the past, and that they may find their widest acceptance on estate crops in the Third World, where regulations are weaker.) He hears scare stories behind what I say, that genetic engineering will bring monsters. (I think, I never said this.) He says he will tell me two stories. When pioneers grew fungus to make antibiotics, everyone said, Oh fungus is going to take over the world, and infest everyone and everything. But anti-biotics saved lives. When the railroads were built, there were articles in this country that it would go eight miles an hour and suffocate people. I can go down through history with these scare stories, he says. I say, but things like that do prove to have problems. Look at the virulent strains of bacteria which have evolved to resist antibiotics. His companion and my acquaintance, Dr T V Sathyamurthy of the politics department at the university, says, he himself evaluates these things in terms of social effects. The railroads for instance: there is a body of literature, very well documented, on how the railroads changed society and, like industrialisation itself, brought problems. Dr Raghaven answers, what you have to do is make a cost-benefits analysis. Antibiotics saved how many lives. He is over 60 years old. He can't believe he is more open-minded than people half his age. You must do experiments and look at the data, he repeats. Sathyamurthy responds, when you show him scientific data, he wants to know the question you asked, and why you asked it. A problem he sees is that scientists have their own world view, social priorities, intellectual leanings. Why you ask a question is a social matter. It is important, the reasons why certain questions are asked, and why other questions remain unasked or ignored. Dr Raghaven throws up his hands: are you telling me you don't trust me or other scientists who give you data? You put hydrogen and oxygen together and get water. You put sodium and chlorine together and get salt. This is straightforward and true. Sathya, my dear and good friend, you and I are chemists . . . (I am too, I interrupt) . . . and you tell me what is the reason why a chemist finds these things out. Sathya says, we are not talking here about those things. We are talking about scientific questions being answered and used that have a social implication, in these cases reproduction and the environment. Sathya continues, he doesn't have a great trust like you do. *You* will not be controlling these things, and your friends will not. The multinationals will. Dr Raghaven quips, he doesn't understand a word we are saying. You see how far behind I am! he jokes. You do controlled, small-scale experiments and see if things work. Then you know. He repeats, there are always scare stories. He says, we are not communicating. He believes the only way to go forward is scientifically. I ask, the scientific temperament? quoting Nehru's often-used phrase, his vision for a secular Indian free of religious strife.

I know a woman, a historian and feminist, who does not believe that genes exist. Although I would not go that far myself, I do not believe the genetic hypothesis – gene-think – is factual in the sense it is purported to be. This is not the same as saying that curiosity about heredity and biology is in itself dangerous. I am thinking specifically about the genetic hypothesis and its tragic failing: that opportunities to really reassess its theories and practices are not taken up. I am thinking, among other things, that the laws which governed the society of Mendel's pea plants cannot provide the laws of human societies. I am thinking of ideology and the belief that genetic engineering is considered, as boasted the bookjacket of *Man Made Life: A Genetic Engineering Primer*, 'the key to understanding not only the future of bio-technology, but the fundamentals of life itself',[2] that genetic engineering will unlock the 'secrets' of our genes. A friend related that on leaving a high-powered, mainstream ethics conference on these matters, a woman participating asked her, 'What are we so afraid of?'

Another friend says that about all the genetics she finds useful is that she was born a human being instead of a frog because of her hereditary endowment from her mother and father. I think of how the scramble of DNA within a fertilised egg cell is considered so crucial now, that which 'makes' us what we are (pace Aldous Huxley, Robert Edwards and Woody Allen who have all identified themselves with a sperm cell, not a fertilised egg. Huxley, brother of Sir Julian and author of Brave New World, wrote a poem about the one sperm that 'survived' to become himself, 'that One was Me'. Edwards, IVF pioneer, alluded to it in telling the story of the first test-tube baby, and referred to himself as 'the One [sperm] that was Me'.[3] Film-maker Woody Allen dressed up as a sperm – Everyman-as-sperm – in Everything you Wanted to Know About Sex . . .)

Science does not have a monopoly on the practice of explaining the world and our lives. Its claims and business are specific. The power of its socio-scientific ideas and the power it is put to are specific. 'Power structures are – pardon the tautology – powerful,' Susan George wrote. 'Otherwise they wouldn't be worth writing or reading about – or acting against – "powerful" does not mean "invulnerable".'[4]

Trying to get to grips with genetic engineering is not a matter of understanding the methods of gene splicing and gene manipulation, but of asking the right questions. Why is everyone being pitched the idea of genes: genes-the-cause, genes as the negotiable entity, 'genes and atoms for health and prosperity'. If these particles called genes are an interesting idea whose time has come to full fruition with genetic engineering, why everywhere you turn are the costs so very steep? Why do future promises that cannot possibly be achieved become the motivating principles while the roots of present and future problems are ignored? Why the duplicity of promising health and prosperity while it is business as usual? Why the over-confidence of its abilities as witnessed in claims of universal success and superiority, and the denials of its obvious limitations, failures, uncertainties, the particularly narrow (not universal) world view and, perhaps most importantly, the inability to recognise when science as a practice is unscientific by its own standards.

A particular case of the latter was exposed as a result of an outbreak of hepatitis among women in the IVF programme at the Dijkzigt Hospital in the Netherlands between 1987 and early 1988. Medical ethicist Helen Bequaert Holmes related the incident. The culture medium which was used to grow eggs and embryos for the 172 women who reached the embryo transfer stage became contaminated with hepatitis B virus. The source of the virus was the human blood serum which was added to the growing medium. Adding blood serum allegedly provides a protein source for the early embryo, but there is no real evidence that it is necessary. Holmes emphasised that, despite this lack of evidence, adding blood serum to the IVF culture medium is a common practice in IVF programmes worldwide. She argued that the ultimate cause of the incident was not just sloppiness and carelessness in the Dutch IVF laboratory, but the 'experimental nature of IVF, the, "hey, what-shall-we-try-next" hocus-pocus, and the gambling, risk-taking behaviour that it tends to foster', which allows well-trained scientists around the world to carry out cavalier trial-and-error experiments.[5]

This trend goes unrecognised; it is as if a general amnesia takes hold or else a general inability to see things that do not fit into the ideals of good scientific practice, or else an ability to turn standard practice into 'aberrations' or 'tragic mistakes'.

Those who justify biological engineering as a way to organise industry, medicine, reproduction and agriculture – in short, living things – often dismiss challenges to it with clichés about steam engines ('people were afraid of the steam engine at first too'), about Galileo ('Scientists on the forefront of discovering the Truth about biology are being persecuted just as Galileo was by the Church'), about progress ('Science and technology mean going forward; resistance to technology means going backwards'), about security and human comfort ('If you don't let us do this, people will starve and remain vulnerable to diseases'). The means and ends – destruction of whole ways of life, great risk, exploitation of living resources and human beings – are not of consequence, or are ignored, or are blamed on something else like a 'population explosion', or the need for better oversight mechanisms to regulate the practice.

The gene revolution is an ideological nexus of commerce, ethics and social relations. Instead of asking, 'What kind of technology do we need to make life better?', the technology – as the manifestation of a particular set of social and economic themes – sets the standard of our meaning of life and our 'needs'. How is it that the promise of genes for a perfect child, a cure for everything, food and energy for all is posited? It is incomprehensible and deeply technocratic. Ah. I think I now understand my friend's disbelief in genes.

Yet we must take genes seriously. All of the attention drawn to genetic engineering, from great enthusiasm to the multi-faceted controversy, returns to a short but provocative statement: genetic engineering will change our landscape and it may even change our species. For this reason and another, namely that it is hard to be well informed on this subject (not because the subject matter is difficult to understand, but because the informers are mainly mouthpieces for science and industry, unable to challenge the assumptions behind genetic engineering), an impoverished gene ethic is one that simply suggests carving up genetic engineering into the beneficial or hazardous applications. The individual and the global effects of genetic engineering implicate our relationship to and in nature. This is one of the reasons, I think, why the phrase *gene ethics*, bold in its demand for our making of moral judgments, has gained so much currency. The other reason is body politics,

something feminism understands very well. The knowledge
– ideas – of genetics and the practice of genetic engineering
is directed at our beings, at our bodies and at our human
identity. We do not, after all, refer so directly to an atom
ethics or chemical ethics.

That said, I am wary of the danger of a critical opposition
to gene manipulation of human beings which is couched in
terms of the special status of our genes. This may sound like
a contradiction. Elsewhere I have argued that the social and
individual harm that is caused by genetic engineering is
significant. Gene splicing and recombination, gene cloning
and cell fusions are significantly unique in that they do alter
the entire genetic make-up directly, in unprecedented and
unpredictable ways. But I would not wish to idealise genes:
to say that the idea that found a physical form, genes, is that
which makes us uniquely human; that which holds the key
to the human condition; that which limits or extends our
possibilities as living creatures, as individuals and as a
species. Reducing humanity to the result of biological –
genetic – determinants is mythical. Feminism understands
this very well, too. Women are not born, they are made.
Organisms are not born, they are made.

However, it is worth considering why many people
(sometimes myself included, I realise) do find gene manipulation of human beings repugnant because of what genes
have come to signify. Here perhaps is a contradiction
because I otherwise find in this attention to genes a
biological determinism, in the service of ideologies, and I
find in it also something very close to the idealisation of the
embryo.

Science and technology have been the vocabulary of our
relationship with nature. We have changed and so has what
we call nature changed in this context. Technology is what
makes the world we know possible, the vocabulary of
perceived possibilities, while outside that vocabulary,
possibilities often seem like pipe dreams, not because they
are really ridiculous but because of the presumption that

technology reflects the history of human progress.

Genes are an issue because the science and technology of genetics, in relationship with biological theories of evolution and life, have made them an issue. Evolution theory says that the mechanism for differentiation of living things is found in the tiny genetic changes – mutations – that amount to big changes over time. How presumptuous is the genetic engineer. It is understandable that an opposition to gene manipulation might be couched in these terms. On the other hand, transformative social movements – all of those discussed in this book – are possibilities beyond these terms.

Vandana Shiva warns of 'green' capitalism and 'eco-imperialism', the belief that an ecological future will be created on the basis of higher rates of economic growth, new life-threatening technologies such as genetic engineering and a renewed control of the world's resources by industrialised countries. She puts the three facets together.[6]

The Penang Declaration was one of many international calls for ecological and sustainable approaches to agriculture which emerged out of a growing concern over the biotechnology 'revolution'. It has a straightforward analysis and recommendation. The long-term costs of biotechnology are too high and its direction morally bereft. Rather than getting at any 'Truth', biotechnology uncovers 'facts of nature amenable to manipulation and control' in common with other present-day technologies. The Declaration called for a different motivation from which research and farming practices can flourish.[7]

When the international information and communication clearing house Gen-ethisches Netzwerk – Gen-Ethic Network – was founded in 1986 in Berlin, to facilitate a public dialogue about the objectives, applications, dangers and consequences of genetic engineering, and to discuss possible alternatives, it questioned the place of the small-scale controlled experiment:

Today it is possible to set into motion global processes

which can neither be reversed nor reliably restrained, and which may end in the annihilation of our planet. It is the ruling paradigm of scientific and technological 'progress' which has brought us to this level of potential destruction – a paradigm which dictates that whatever can be done *must* be done, and that it is only in the application of new knowledge that the consequences become evident. This notion of trial-and-error is clearly no longer adequate to the dimension of the problem we face today. We must develop a new ethic for dealing with our knowledge and cannot entrust this task only to scientists, politicians, and so-called experts, nor can we leave it to the mechanisms of the free market and international competition.[8]

The call for a new ethic sounds unpatriotic to the science which enforces industrial principles, but, on the other hand, there is a terrific tension among those working in science about the realities of genetic engineering. At the very least, the new ethic is necessary for:

- an honest reappraisal of the harm, not only the harm biological manipulation directly causes and may cause, but in terms of the neglect of people's real needs when worthless so-called solutions are sought
- an honest evaluation of the actual feasibility and benefits of various biotechnologies, and of who disproportionately benefits and suffers; as we have seen there is a huge reality gap in the abstract notions, such as higher-yielding crops, perfect pharmaceuticals and healthy babies
- a reorientation of priorities, to address the question, 'What transformations in our lives will enhance the welfare of all human beings?' rather than the question, 'How can we apply genetic engineering?'
- an examination of the shift in meaning of eugenics, its meaning beyond the dictionary definition of 'race

improvement'; there seem to be as many different meanings as there are individuals (including myself) and pressure groups who use the term, which means that gene ethics come in many varieties. After all, Robert Sinsheimer, a vocal protester against the militarisation of biotechnology, thought up the human genome project, at worst a eugenicist dream come true, at best fraught with struggles against forces of discrimination and oppression, anticipating the need for protective measures to maintain human rights. This project may be in its effects on an individual, group or population as cruel and disrespectful as a biohazard.

After such re-examinations, can there be an ethical genetic engineering? In the sense of the bio-revolution, I think not. In the sense of an isolated use, probably, but many isolated instances are measured in a vaccuum, and the lack of harm of an isolated case is not the point. The benefit is. (I know elements in the complete history of the petro chemical and nuclear age may be recorded as triumphs. Plastic disposable (sterile) sryinges, for instance. Chemotherapy, for instance.)

What all this means in terms of ethical limits is the more difficult question, since we already live with so much genetic technology, since the boundaries between the 'old' and the 'new' are not definite, since every human activity is bound up with so many others, and since knowledge is the sum of many endeavours and experiences. But knowledge too can be re-examined and redirected.

In the end, gene ethics cannot only be about restraint – restraining the technology. It is necessary, but not sufficient. Gene ethics is not aimed at deprivation, but fulfilment: sustainable development, transformative feminism, environmental welfare, humane treatment of animals and all living things, in short the more just world which these point to . . .

Speaking of the challenges of new biotechnologies at the historic Bogève meeting, Tim Brodhead, Executive Director of the Canadian Council for International Cooperation, said, 'We stand at a moment of profound change, in which the values of a production-centred model of development compete with those of people-centred development, and Western technology-driven progress with a new paradigm which is both more sustainable and more tolerant of cultural particularities. Tomorrow has no fixed address; the role of people's organisations, as always, is to champion pluralism, the capacity of individuals to create the future they wish for themselves and their children.[9]

I wonder, if all the resources and human energy and creativity that have already been sunk into the bio-revolution, defined as the capability to alter the heredity and characteristics of living things, instead were directed toward a clearer understanding of the economic and social forces (including scientific, technological ones) which cause so much human suffering, and towards the answers to human welfare needs – clean air, for instance – what kind of questions would a science be asking, what kind of tech-nologies would be forthcoming and what would a social revolution served by science – a science for people – be?

Appendix   Gene Therapy

Gene therapy is gene manipulation done on human cells, aimed at the cells of human adults, children, foetuses or embryos. It is genetic engineering by another name. It is difficult to describe without falling into the language of 'defective genes', since its premise is a rather mechanical understanding of the relationship of genes and illness, as if fixing a gene, like fixing a faulty spark plug in a car, is going to solve the basic problem of a genetic ailment. And these days, that means many more ailments than ever were considered genetic before, as expressed in the suggestion, ' "Gene therapy" to replace a missing gene may one day cure many forms of cancer.'[1]

Gene therapies entail many different approaches. Gene *replacement* therapy is the addition of genes to human cells; it would be used if the disorder is linked to the *lack* of a working gene. For disorders like Huntington's chorea which are thought to be due to the *presence* of a particular gene, gene therapy would mean *deactivating* the gene in question. Gene *activation* therapy aims to activate a normally dormant gene which is thought to have a similar function to a 'faulty' gene.

Gene therapy implies a medical reason behind its use, but it is not a straightforward concept. Peter Wheale and Ruth McNally point out in *Genetic Engineering: Catastrophe or Utopia?*

characteristics (for example adding the gene which codes for growth hormone with the aim of making a child grow taller), or as an adjunct to therapy. In fact the first permitted clinical use of gene therapy in the USA, in which genetically engineered cells were injected into a human patient for the first time, had nothing to do with the original definition of gene therapy, 'the insertion of a normal gene which then corrects a genetic defect'.[2] Rather, genes were used as markers to follow the course of a non-genetic cancer therapy. The cancer therapy uses a type of white blood cell, known as a tumour-infiltrating cell, which shows promise for therapy because of its ability to 'attack' certain cancer tumours. Researchers added a gene to tumour-infiltrating cells as a marker in order to follow their course in the body of the patient. Thus they could judge the merits of using tumour-infiltrating cells for cancer treatment.

State of the Art

There are several approaches available to alter the genetic make-up of human cells. In principle any of the methods already used on animal cells could be used on human cells. One of the earliest methods experimented with entails inserting foreign genetic material into an individual's bone marrow cells. It might be carried out as follows:

1 Isolate a gene of interest from human DNA. Insert the gene into retroviruses where it will be incorporated into the viral genetic material. Or, as in the case of gene therapy directed at the liver, hepatitis virus is being used instead of retrovirus. Other pieces of DNA, called promoters and enhancers, must be used to create the conditions for the gene to be incorporated into other genetic material.

2 Remove bone marrow from the human subject.

3 'Transfect' (infect) the bone marrow cells with the retrovirus carrying the gene of interest.

4 If the bone marrow cells are successfully transfected, and if the new gene appears to be working properly in the cells, then replace the patient's bone marrow. If all goes as planned, the new gene will work within the person's body.

Bone marrow cells are used because the procedure requires some source of cells which rapidly divide (why this is necessary is unknown), and because they can be removed from the body and replaced. A small percentage of the bone marrow cells are stem cells, and these are the cells which must become infected with the new DNA. Stem cells are precursors to red blood cells and other important cells. They are 'immortal' in that they continue to produce cells in the bone marrow indefinitely. Since the most common and most studied inherited genetic disorders are of the red blood cells (such as sickle cell anaemia and the thalassaemias), this approach was considered the most promising at first.

Somatic Cell and Germ Cell Therapy

Some medical scientists and ethicists divide gene therapy into two distinct categories, somatic cell and germ cell manipulation. Somatic cells are all the body's cells other than the reproductive cells; thus a somatic cell mutation is one that is not heritable. The bone marrow approach is an example of somatic cell therapy, in that only certain body cells are affected, but not reproductive cells (apparently).

The germ cells are those in the series of cells called the germ line, which eventually produce the reproductive cells, the egg and sperm. Egg and sperm cells each carry half a complete set of genetic material. In mammals, the ovaries and testes contain the germ cells.

Germ cell or *germ line* therapy is gene manipulation of the fertilised egg (early embryo). Germ line gene manipulation is already done on animals for research, farming or industrial purposes. The prospective purpose in gene therapy is to manipulate the genome of the human embryo so as to avoid the possibility of a genetic disease. If such a manipulation were to succeed, the genetic change would be carried in every cell of the resulting person's body, including the reproductive cells. The genetic alteration would be heritable, and passed on to subsequent generations.

IVF or *in vitro* fertilisation opened the door to the possibility of manipulating the fertilised egg (since it must be done outside the woman's body), and that possibility is one of the most controversial ethical issues of bioengineering and new reproductive technology. Germ cell genetic engineering is considered the more ethically complex operation, since it would alter the genetic make-up of future generations, thus altering the course of human (genetic) evolution – a social experiment of novel proportions.

Still, the distinction between body cell and germ cell therapy is not as clear-cut as it is made out to be. For one, the argument has been made that if somatic therapy works, then the people treated thus will reproduce with their 'faulty genes'. So you should 'fix' their egg and sperm cells to prevent this. Another argument is made that if gene therapy is accepted, then some diseases may only be treatable by the germ cell approach because of the brain/blood barrier (that is, brain cells may be inaccessible to somatic gene manipulation). For these reasons, and probably more, many advocates of gene therapy have argued that we should not be too hasty about rejecting germ line therapy. Who knows what good it could do, they say.

In their evidence to the British government's Committee on the Ethics of Gene Therapy, the public interest group Consumers for Ethics in Research (CERES) questioned the way in which somatic cell gene therapy and germ cell gene therapy were considered two separate categories when it came to drawing up regulations and ethical guidelines.[3] The government committee said they were only interested in somatic cell therapy; germ line therapy was banned and so was not at issue. But any recommendation on somatic gene therapy, CERES said, will inevitably influence future discussion of germ line gene techniques, just as today advocates of gene therapies refer back to debates about organ transplants to suggest that gene therapy is nothing new. (The argument is that gene therapy is simply bone marrow transplantation with an extra step. But what an extra step.)

Claims of Benefit

At first gene therapy promised to be a dramatic cure for the simplest of the inherited disorders, those called 'single gene disorders' such as haemophilia and thalassaemia. If the single gene is identified and isolated, it could be incorporated into the genetic material of a large number of cells in the person's body, where hopefully it would act perfectly well and cure the disorder. In 1985 big successes were predicted for gene therapy treatment of the rare Lesch Nyhan syndrome and ADA (adenosine deaminase) deficiency (also called Severe Combined Immune Deficiency, the illness that 'the bubble boy' suffered from). ADA can be treated, apparently reasonably successfully, by bone marrow transplant. But ADA was considered an ideal candidate for gene therapy. It seemed more a case of using the ailment to perfect the therapy than for actually creating a far superior therapy for ADA itself. In 1986, despite known problems with the gene therapy technique, eight or ten laboratories in the world were pursuing gene therapy for Lesch Nyhan syndrome and ADA deficiency. Interest continued to grow in it, which raised the bleak comment that there are more people working on a genetic cure for ADA deficiency than there are people who have it.

Researchers were having a difficult time of getting gene therapy to work on the red blood cell disorders. Gail Vines, writing in *New Scientist* in 1986, said that researchers were therefore looking for other diseases to treat by gene therapy.[4] The situation demonstrates a remarkable inversion of priorities. Instead of looking for a therapy for a specific disease or illness or disability, researchers look for an ailment for the therapy.

Some Problems:

By conservative estimates, questions of effectiveness and safety arose.

- The extraneous bits of DNA which are incorporated with the gene, called promoters and enhancers, may suppress the expression of the gene of interest. The body may not see it as a functional gene.
- Even if the foreign DNA is inserted into the person's cells successfully, the complex processes that are involved are unknown. This became apparent in early research on hereditary anaemias, the haemoglobin disorders. By 1986, when it was clear that the gene therapy for them was not as straightforward as was predicted, researchers explained that it was because production of haemoglobin is so complicated. There are a number of biochemical reactions going on and a number of genes involved. These were out of reach for gene therapy at the time, they said. (We have come full circle. In 1988 researchers were reporting success with controlling the regulation of these genes. I suspect that these disorders have reappeared on the agenda for gene therapy.)
- Insertion of the new DNA into the human genome is by definition a gene mutation and hence could be carcinogenic (which makes me wonder about the prediction

that it will one day cure cancers).

- Similarly, when new genetic material is inserted into the individual's DNA, it does not incorporate itself in a specific position in the DNA; instead, the result is a random insertion. Ideally, there should be a specific insertion at a specific site in the DNA, namely where the gene would normally occur. Random insertion possibly will cause problems, as has occurred in animal experiments. (A pig carrying the human growth hormone gene suffered arthritis, heart disease and metabolic effects, although how much such effects may be due to random insertion and how much to other factors introduced with the gene insertion method is not clear.)
- Using viruses to shuttle genes into cells is risky, for instance the weakened virus may recombine with undetected viruses or DNA sequences in the cells and so become infectious.
- As with other experimental therapies, questions arise as to how patients are selected; how informed consent is acquired; as to privacy and confidentiality in contrast to the public right to know, etc.
- The earliest patients of gene therapy will be infants and young children, since genetic disorders account for many of those in hospital; as research subjects, they generally are in a more vulnerable position than adults.

Ethics

The attitude 'it must be pursued' dominated the news on gene therapy. The early in-house medical science ethics discussion took its cue from the perceived technical problems, such as, will the cure be worse than the disease? There was little or no discussion of the wider issues of health care provision, the ethics of using human beings – children and adults – as experimental subjects, the social implications of identifying genes as the cause of ill health. Nor was the changing philosophy of cause and prevention addressed (when genes are the cause, genetic therapy is the solution), nor the high-tech bias in the use of genetic engineering for therapy. There seems to be little discussion about the changing direction of clinical medicine with genetic engineering generally, but shifting directions there are. In England, powerful advocates of the new genetics in clinical medicine have achieved some institutional changes. An Institute of Molecular Medicine replaced the former Nuffield Institute at Oxford with the purpose of organisationally applying molecular biology to medicine. One of its principal planners, Sir David Weatherall, is a staunch advocate of the new genetics in medicine, including gene therapies.

In the USA, unauthorised gene therapy experiments were carried out on two patients with thalassaemia by Martin Cline at the University of California at Los Angeles in 1980. It prompted some agreement that general ground rules must be applied to gene therapy experiments.

Wheale and McNally argue against somatic gene therapy on straightforward technology assessment and health economics grounds. They conclude that any such treatment is unlikely to work well; and that the consequential benefits to the patients are unlikely to be great. Financial and social costs and risks which will arise for both patients and the wider community make it unlikely to increase social welfare.[5]

Other questions arose regarding altering the fertilised egg cell and thus the germ line of the resulting individual, ultimately evolution and, as some suggest, the species human beings may be. Is it moral to allow scientists to make such alterations to the fundamental characteristics of individual human beings and future human beings? In this sense, are germ lines special and should they be out of the bounds of genetic engineers? Such questions are not limited to the medical scene, no matter how broadly genetic medical conditions might be defined; they are extended to the possibilities of using biological engineering to create people to order, the dystopia of a Brave New World.

Many critical analysts argue along a moral line against manipulating the germ cells of human beings on the grounds that it is tampering with fundamental human characteristics. Whether one accepts that human germ lines are sacred or not, for women, the implications of such a level of genetic engineering are unique; as fertilised eggs (embryos) and foetuses become the focus of more therapies – gene manipulation among them – women become the medical and experimental subject.

Notes

In the following Notes, the details of the volume and number of a journal reference are given as two figures, eg 1 (1) means volume 1, number 1.

Introduction

1 E P Thompson, *The Making of the English Working Class*, Penguin, Harmondsworth, 1980, pp. 570, 601–4.

2 Susan George, *A Fate Worse Than Debt*, Penguin, Harmondsworth, 1988, p. 5.

3 Cary Fowler, Eva Lachkovics, Pat Mooney and Hope Shand, 'The Laws of Life: Another Development and the New Biotechnologies', *development dialogue*, 1–2, 1988, p.289.

4 Tim Brodhead, 'Tomorrow Has No Fixed Address', *development dialogue*, 1–2, 1988, p. 297.

1 The Gene Revolution

1 Fowler, Lachkovics, Mooney and Shand, 'The Laws of Life, p. 25.

2 Report of the Committee of Enquiry on Prospects and Risks of Genetic Engineering, January 1987, Deutscher Bundestag: III; translation by Professional Translators, Dunstable, England.

3 Ruth Jarvis, 'Global Warmer', book review of *The End of Nature* by Bill McKibben, Viking, London, 1990 in *Books*, Waterstones, February, 1990, p. 8.

4 Carol Ezzell, 'Congress Provides Rundown on US Biotechnology', *Nature*, 28 July 1988, p. 283.

5 Peter Marsh, 'Why ICI Believes People Need Space', *Financial Times*, 9 August 1989, p. 12.

6 Sir Geoffrey Allen, 'Leverhulme Lecture: Our Chemical Industry in 2001', *Chemistry & Industry* 11, 4 June 1990, p. 353.

7 Picture reproduced for a review of Douglas Dixon's *Man After Man: An Anthropology of the Future*, Cassell, London, 1990.

8 J Coombs, *The International Directory of Biotechnology 1985*, Macmillan, London, 1985, p. xi.

9 *World Resources 1988–89*, World Resources Institute and the International Institute for Environment and Development in collaboration with the United Nation's Environment Programme, Basic Books, New York, 1988, p. 51.

10 Bernhard Claußen, 'Modern Genetic Engineering as a Political Issue of the Contemporary Risk-fraught Society', *Proceedings of the European Workshop on Law and Genetic Engineering, Hamburg, 14–15 December 1989*, Hamburg, 1990, p. 8.

11 Anne McClintock, 'Dangerous Liaisons', book review in *The Women's Review of Books*, May 1980, p. 4.

12 Peter Wheale and Ruth McNally, *Genetic Engineering: Catastrophe or Utopia?*, Harvester Wheatsheaf, England, 1988; Jean L Marx, (ed.), *A Revolution in Biotechnology*, Cambridge University Press, Cambridge, 1989.

13 See Grahame Bulfield, 'Genetic Manipulation of Farm and Laboratory Animals', in, Peter Wheale and Ruth McNally, (eds.) *The Bio-Revolution – Cornucopia or Pandora's Box?*, Pluto Press, London, 1990, p. 18.

14 Edmund W Sinnott, L C Dunn and Theodosius Dobzhansky, *Principles of Genetics* (fifth international student edition), McGraw-Hill, 1958, p. 36.

15 Royal Commission on Environmental Pollution (RCEP), *Thirteenth Report: The Release of Genetically Engineered Organisms to the Environment*, Cm 720, HMSO, July 1989, p. 13.

16 Wheale and McNally, *Genetic Engineering*, p. 28.

17 Mary Midgley, 'Forward thinking', the *Guardian*, 2 December 1987.

18 Germaine Greer, *The Female Eunuch*, Paladin Books, London, 1971, p. 25.

19 Tim Wilkie, 'Mice Embryos' Sex Changed', the *Independent*, 9 May 1991.

2 Biotechnology Now

1 Promotional letter for the international monthly for industrial

2 Wolfgang Goldhorn, speaking on 'The Welfare Implications of BST' at the conference Action Alert: The Bio-Revolution, Cornucopia or Pandora's Box?, London, 7–8 October 1988; presentation published in Wheale and McNally, *The Bio-Revolution*.

3 RCEP *Thirteenth Report*, p. 10; Stephen Oliver, 'Prospective Uses', a book review, *Times Higher Education Supplement*, 7 July 1989.

4 Lancelot Hogben, *The Retreat from Reason*, Watts & Co., London, 1936; Introduction by Julian Huxley, p. vii.

5 Bernard Dixon, 'Bug Bang that Built Israel', *Observer Magazine*, 1 November 1987, p. 19.

6 Stephanie Yanchinski, *Setting Genes to Work: The Industrial Era of Biotechnology*, Penguin, Harmondsworth, 1985, pp. 125, 127.

7 Eric Brunner, *Bovine Somatotropin: a Product in Search of a Market*, London Food Commission, 1988; private conversation with Eric Brunner, 9 September 1988; *Dispatch*, 'Milky Business', Channel 4, 24 June 1988; Dorothy Wade, 'A Country Flowing with Milk and Drugs', the *Guardian*, 10 July 1988.

8 Wade, 'A Country Flowing with Milk and Drugs'.

9 Richard Lacey, 'Milk of Unkindness', the *Guardian*, 6 October 1989; James Erlichman, 'Supermarket Chains Demand End to Secrecy over Addition of BST Hormone to Milk Supply', the *Guardian*, 22 July 1988; Paul Brown, 'Scientists "Gagged over BST Milk Risk"', the *Guardian*, 8 August 1989.

10 Samuel S Epstein, 'Potential Public Health Hazards of Biosynthetic Milk Hormones', the *Ecologist*, September/October 1989, p. 1.

11 James Erlichman, 'Taking a Rise out of "Mutant" Bread', the *Guardian*, 3 March 1989.

12 Epstein, 'Potential Public Health Hazards ...'; Brunner, *Bovine Somatotropin*; Rainbow Group GRAEL (European Parliament), 'How Do You Like Your Milk?', information leaflet on BST, undated, distributed 1989.

13 H Patricia Hynes, 'Biotechnology in Agriculture; An Analysis of Selected Technologies and Policy in the United States', *Reproductive and Genetic Engineering: Journal of International Feminist Analysis*, 2(1), 1989, pp. 39–49, 41.

14 The conference proceedings were published in Wheale and McNally's *The Bio-Revolution*.

15 GRAEL, 'How Do You Like Your Milk?'

16 Pat Roy Mooney, 'The Law of the Seed', *development dialogue*, 1–2, 1983.

17 John Webster, 'Sense and Sensibility Down on the Farm', *New Scientist*, 21 July 1988, pp. 41–4, 41.

18 European Parliament, extracts from European Parliament's Working Document Report on behalf of the Committee of Agriculture, Fisheries & Food, on the effects of the use of biotechnology on the European farming industry. Rapporteur: Mr F W Graefe zu Baringdorf, Document A 2–159/86, 26 November 1986.

19 GRAEL, 'How Do You Like Your Milk?'

20 Vandana Shiva, *Staying Alive: Women, Ecology and Development*, Zed Books, London, 1988.

21 Steven Dickman, 'Next Round in West Germany's Biotechnology Licensing Struggle', *Nature*, 15 September 1988, p. 199.

22 Felicity Lowe, lette to the editor, the *Guardian*, 4 August 1989. With reference to Aileen Ballantyne, 'Insulin Warning After Doctor Dies', the *Guardian*, 29 July, 1989; see also Frank Lesser, 'Human Insulin Comes Under Scrutiny as Number of Deaths Rises', *New Scientist*, 19 August 1989, p. 22.

23 Edwin Gale, 'Hypoglycaemia and Human Insulin', the *Lancet*, 25 November 1989, pp. 1264–5, 1266.

24 Nicholas Russell, 'Slow Progress in the War on Diabetes', the *Independent*, 9 October 1989.

25 Stephen S Hall, *Invisible Frontiers: The Race to Synthesize a Human Gene*, Sidgwick & Jackson, London, 1988.

26 *Ibid.*, p. 302.

27 See, Robert L Peters II, 'The Effect of Global Climatic Change on Natural Communities', in Edmund O Wilson, (ed.) and Frances M Peter (associate ed.), *Biodiversity*, National Academy Press, Washington DC, 1988, p. 451.

3

1 Panel, Women's Hearing on Genetic Engineering and Reproductive Technologies at the European Parliament, Brussels, 6–7 March 1986.

2 See Berg et al., *Science* 185, 1974, p. 303.

3 Wheale and McNally, *Genetic Engineering*, p. 140.

4 Carol Ezzell, 'Unauthorized Environmental Release Cleared by NIH', *Nature* 331, 1988, p. 202.

5 Quoted in Charles Piller and Keith R Yamamoto, *Gene Wars: Military Control Over the New Genetic Technologies*, Beech Tree Books, New York, 1988, p. 195.

6 For a review of the different regulations in the US and UK see Wheale and McNally, *Genetic Engineering*.

7 *New Scientist*, 'Cancer at the Pasteur', 18 June 1987, p. 29.

8 John Elkington, 'Natural Attacks', the *Guardian*, 31 May 1988.

9 *European Chemical News*, 'C-G Rejects Seed Herbicide Package', 21 November 1988, p. 26; The Genetics Forum information leaflet on Genetic Engineering and Food, distributed 1990; Mooney, 'The Law of the Seed', p. 102.

10 *European Chemical News*, 'Ciba-Geigy Buys Chiron Stake', 21 November 1988, p. 26.

11 From my conversations with Professor John Lawton, Department of Biology, University of York.

12 RCEP, *Thirteenth Report*, p. 20.

13 Quoted in Piller and Yamamoto, *Gene Wars*, p. 188.

14 Wheale and McNally, *Genetic Engineering*, pp. 178–9.

15 Marcia Barinaga, 'Field Test of Ice-minus Bacteria Goes Ahead Despite Vandals', *Nature*, 30 April 1987, p. 819.

16 RCEP, *Thirteenth Report*, p. 19; *geneWATCH*, 'Environmental Release of Genetically Engineered Organisms: Recasting the Debate', March/June 1988.

17 Quoted in Dick Russell, 'Wrong Message Sent on Genetic Engineering', the *Guardian Weekly*, 9 September 1987, p. 5.

18 Paul Hatchwell, 'Opening Pandora's Box: The Risks of Genetically Engineered Organisms', the *Ecologist*, 14(4), 1989, pp. 130–6.

19 See David Bishop, 'Genetically Engineered Insecticides', in Wheale and McNally, *The Bio-Revolution*.

20 News release, 'HSE SHOULD RETAIN RESPONSIBILITY FOR REGULATING GENETIC ENGINEERING – CBI', Confederation of British Industry, 30 August 1989. (Source: The Genetics Forum, London.)

21 Quoted in Jon Turney, 'Fears Rise on Genetic Engineering Safety', *Times Higher Education Supplement*, 28 December 1988.

22 William Freudenburg, 'Perceived Risk, Real Risk: Social Science and the Art of Probabilistic Risk Assessment', *Science*, 7 October 1988, pp. 44–9.

23 Mark Dodgson, 'What Future for Biotech Firms?', *Chemistry & Industry*, 4 June 1990, p. 368. Genentech, the most successful small biotechnology firm in the world, sold a majority shareholding to the well-established Hoffman La-Roche corporation.

24 'Flesh Ties', an interview with Christa Wolf by Aafke Steenhuis, *New Statesman and Society*, 23 February 1990, pp. 26–30.

25 See RCEP, *Thirteenth Report*, Steve Connor, 'Genes on the Loose', *New Scientist*, 26 May 1988, pp. 65–8.

26 Tim Cooper, 'Whole Earth Catastrophe', the *Guardian*, 12 May 1986.

4

1 Susan George, *How the Other Half Dies: The Real Reasons for World Hunger*, Penguin, Harmondsworth, 1976, p. 89.

2 George, *A Fate Worse Than Debt*, pp. 14, 15.

3 Mooney, 'The Law of the Seed', p. 95.

4 *Ibid*.

5 Calestous Juma, *The Gene Hunters: Biotechnology and the Scramble for Seeds*, Zed Books, London, 1989, pp. 52ff discusses plant exchanges, and the sexual and racial exploitation these engendered.

6 Donald Plucknett, et al., *Gene Banks and the World's Food*, Princeton University Press, Princeton, NJ, 1987, p. 58.

7 Mooney, 'The Law of the Seed', ICDA Seeds Campaign, see Resources.

8 See Juma, *The Gene Hunters*; George, *How the Other Half Dies*.

9 See Mooney, 'The law of the Seed', p. 86.

10 Omar Sattaur, 'The Shrinking Gene Pool', *New Scientist*, 29 July 1989, pp. 37–41.

11 *Ibid*.

12 George, *How the Other Half Dies*, p. 89.

13 *Ibid*., p. 132.

14 Mooney, 'The Law of the Seed'.

15 *Ibid*.

16 *Ibid*., p. 29; see Plucknett, et al., *Gene Banks*, pp. 189ff.

17 Fowler, Lachkovics, Mooney and Shand, 'The Laws of Life', p. 257–8.

18 European Parliament, Document A 2–159/86, op. cit., p. 14.

19 Mooney, 'The Law of the Seed', p. 19.

20 Statistics from Professor Gordon Goodman, Beijer Institute, World Commission on Environment and Development, 'Our Common Future', a talk and paper presented at the University of York, 1987.

21 Adrienne Rich, *The Dream of a Common Language: Poems 1974–1977*, Norton, New York, 1978.

22 Goodman, 'Our Common Future'.

5

1 *Patenting Life Forms in Europe, Proceedings*, An International Conference at the European Parliament, 7–8 February 1989, ICDA Seeds Campaign, Barcelona, March 1989, p. 64.

2 Animal Biotechnology Cambridge, *Research Into Profit: Aims and Practices of the Animal Research Station*; the *Guardian*, 14 September 1988, publicity photograph and caption.

3 Robin McKie, 'Clone Cows for Farms', the *Observer*, 14 February 1988.

4 See Mooney, 'The Law of the Seed', pp. 151, 162, where he argues that allowing the patenting of plant varieties did not work to increase innovation and profit or to stimulate economic growth.

5 *Ibid.*, p. 137.

6 Quoted in Harriet A Zuckerman, 'Introduction: Intellectual Property and Diverse Rights of Ownership in Science', reprinted in *Current Contents*, 26 June 1989, pp. 4–9, 5. Original in *Science, Technology, & Human Values* 13(1&2), 1988, pp. 7–16.

7 Dr J G Boonman, 'Plant Patenting as Seen by a Plant Breeding Professional', in *Patenting Life Forms in Europe, Proceedings*, p. 30.

8 Benedikt Haerlin, Member of the European Parliament and co-founder of the Gen-Ethic Network, related this at the 'Action Alert' conference, London, 1988. See Wheale and McNally, *The Bio-Revolution*. The analysis in this chapter on patenting life forms in Europe comes from many diffrent sources. See note 12.

9 Mooney, 'The Law of the Seed', pp. 147–8.

10 *Ibid.*, p. 142.

11 Philip W Grubb, *Patents for Chemists*, Clarendon Press, Oxford, 1982, p. 45.

12 Juma, *The Gene Hunters*, p. 162; Grubb, in *Patents for Chemists*, says the British Patent Office granted a corresponding patent much earlier, in 1976.

13 Ruth Hubbard and Sheldon Krimsky, 'The Patented Mouse', *geneWATCH*, January/February 1988, pp. 6–7, 7.

14 Zuckerman, 'Intellectual Property', p. 7.

15 Much of the critique of the European Directive which I present, especially regarding the use of language by the European Commission, is taken from *Patenting Life Forms in Europe, Proceedings*; and *idoc internazionale*, number 2, special issue: 'Biotechnolgy – Where to Now?', March/April 1988.

16 ICDA Seeds Campaign, 'The Patenting of Life: 7 Questions, 7 Answers', *idoc internazionale*, 'Biotechnology', p. 14.

17 Henk Hobbelink, 'Patenting of Life Forms: How to React', *idoc internazionale*, 'Biotechnology', p. 12.

18 ICDA Seeds Campaign, 'The Patenting of Life', p. 17.

19 Irving Kayton, "Does Copyright Law Apply to Genetically Engineered Cells?', *Trends in Biotechnology* 1(1), 1983, pp. 2–3. Adapted from his article in *George Washington Law Review* 50, 1982, pp. 191–218.

20 BBC, 'The Book of Man', *Horizon* series, from the script of the programme, transmitted 9 January 1989 & 10 January 1989, p. 13.

21 Discussed in Zuckerman, 'Intellectual Property'.

22 Sandra Keegan, 'The Proposed Directive on the Legal Protection of Biotechnological Inventions', in, *Patenting Life Forms in Europe, Proceedings*, pp. 10–14.

23 Lori B Andrews, 'My Body, My Property', *Hastings Centre Report*, October 1986, pp. 28–38; for a critique of her position see Maria Mies, 'From the Individual to the Dividual: In the Supermarket of "Reproductive Alternatives"', *Reproductive and Genetic Engineering: Journal of International Feminist Analysis* 1(3), 1988, pp. 225–38; see also Gena Corea, *The Mother Machine*, The Women's Press, London, 1988.

24 Martin Evans, 'Donors and Do Nots', the *Guardian*, 1 February 1989; Aileen Ballantyne, 'Kidney Sale Clinic Bans Two Doctors', the *Guardian*, 8 February 1989.

25 My private conversation with Dr Glatt, when we appeared together as guests on *Kilroy*, BBC 1, 12 May 1989.

26 Marie-Angèle Hermitte, 'Patenting Life Forms: The Legal Environment', *Patenting Life Forms in Europe, Proceedings*, p. 15.

6

1 Frédérique Apffel Marglin, 'Smallpox in Two Systems of Knowledge', in F A Marglin and S A Marglin (eds.) *Dominating Knowledge: Development, Culture and Resistance*, Oxford University Press, Oxford, 1990, p. 123.

2 Tim Beardsley, 'Hepatitis Vaccine Wins Approval', *Nature*, 31 July 1986, p. 396.

3 See Robert Sharpe, 'A Stab in the Dark', *Liberator*, January/February 1989, pp. 12–16; Walene James, *Immunization: The Reality Behind the Myth*, Bergin & Garvey, MA, USA, 1988.

4 James, *Immunization*, p. 33.

5 See John Webster, 'Animal Welfare and Genetic Engineering', in Wheale and McNally, *The Bio-Revolution*, p. 28.

6 Ed Harriman, 'The Good Old British Jab', *New Statesman and Society*, 23 September 1988, pp. 10–12; A H Griffith, 'Permanent Brain Damage and Pertussis Vaccination: Is the End of the Saga in Sight?', *Vaccine* 7, June 1989, pp. 199–210.

7 Ad Hoc Group for the Study of Pertussis Vaccines, 'Placebo-Controlled Acellular Pertussis Vaccines in Sweden – Protective Efficacy and Adverse Events', the *Lancet*, 30 April 1988, pp. 955–60.

8 Nagrik Mahamari Janch Samiti, *Crime Goes Unpunished: A Report on the Cholera Epidemic*, New Delhi, October 1988. See Bibliography.

9 *Ibid*, pp. 36–7, 45.

10 Cited in Barbara Crossette, 'Child Mortality Linked to

Nations' Debts', *International Herald Tribune*, 21 December 1988. Lloyd Timberlake and Laura Thomas reported in 'Death by Adolescence', the *Guardian*, 25 May 1990: 'Every single day, 11,000 children under the age of five in the developing world die of diarrhoeal diseases, most of these caused by bad water. On any given day, 150 million under-fives are malnourished, a condition that leaves them vulnerable to the host of diseases like measles and malaria which kill 14 million under-fives each year. Malnutrition has many causes, many of them related to the environment, including the rapid spread of deserts and the loss of forests and trees, which mean that families can grow less food and have less firewood with which to cook what they do grow.'

11 Shila Rani Kaur, 'Reproductive Technology, Fertility Control and Women's Health: A Third World Perspective', paper distributed at the UBINIG/FINRRAGE conference, Comilla, Bangladesh, 18–25 March 1989.

12 Inserts from Women's Global Network on Reproductive Rights Newsletter, January/March 1989.

13 Malini Karkal, *Can Family Planning Solve Population Problem?*, Stree Uvach Publication, March 1989. To obtain write to 4, Dhake Colony, Bombay 400 058.

14 Quoted in James, *Immunization*, p. 46.

15 Barrie Penrose, 'Smallpox Virus to be Destroyed', *Sunday Times*, 8 October 1989.

16 Sharpe, 'A Stab in the Dark', p. 16.

17 See, for example, H Patricia Hynes, 'Lead Contamination: a Case of "Protectionism" and the Neglect of Women', in, H Patricia Hynes (ed.), *Reconstructing Babylon*, Earthscan, London, 1989.

18 Tracy L Gustafson, et al., 'Measles Outbreak in a Fully Immunized Secondary-School Population', *New England Journal of Medicine*, 316(13), pp. 771–4; related correspondence pp. 834–6.

19 Frank Fenner, 'The Eradication of Smallpox', *impact of science on society*, No. 150, 1988, pp. 147–58, (published by UNESCO).

20 Cited in James, *Immunization*.

21 Marglin, 'Smallpox'.

22 Editorial, *impact of science on society*, p. 105.

23 James, *Immunization*, p. 123.

24 Patricia Spallone, 'The Politics of Vaccination', *Vogue*, London, April 1990.

25 Wheale and McNally, *Genetic Engineering*, p. 179.

26 MRF (Malte Rauch Filmproduktion), *Monkey Business: The Myth of the African Origin of AIDS*, Frankfurt am Main, 1989. Transcript, English translation, p. 35.

27 *The AIDS Catch* (Meditel Productions for Channel 4), documentary on theory of a viral cause of AIDS, aired 13 June 1990.

A reading list which includes works which both support and criticise the claims was supplied by Channel 4, 60 Charlotte Street, London W1P 2AX.

28 Susan Sontag, AIDS and its Metaphors, Penguin, Harmondsworth, 1988, pp. 50, 52.

29 The theory that AIDS originated by animal experimentation is thoroughly treated in National Anti-Vivisection Society, Biohazard: The Silent Threat from Biomedical Research and the Creation of AIDS, NAVS, 51 Harley St, London W1N 1DD, 1987; the laboratory origin theory is discussed in MRF (Malte Rauch Filmproduktion), Monkey Business. The British press reported claims of a germ warfare experiment origin of the AIDS virus, but according to Alastair Hay of University of Leeds who investigated the claim, there is no evidence to support it.

30 MRF (Malte Rauch Filmproduktion), Monkey Business, p. 18.

31 Ibid., p. 24.

32 Contraception: An International Journal 34 (1), July 1986, p. 14–15.

33 Ana dos Reis, 'WHO?', Reader, Second National Congress, Women Against Gene and Reproductive Technologies, Frankfurt, 28–30 October 1988.

34 According to Mr Surish Chandran, public relations officer at the National Institute of Immunology (NII), Delhi. The following discussion is the outcome of my visit to NII, April 1989, and my conversations with various people, including the director, Professor G P Talwar; Mr Chandran; Research Officer Dr Indira Kharat; and Dr Charanjit Babra, WHO consultant visiting NII at the time.

35 Papers reprinted in NII Annual Report, April 1987 to March 1988. See especially, G P Talwar, et al., Clinical Trials with an anti-gonadotropin vaccine, pp. 13–16.

36 My conversation with Professor Talwar.

37 I heard this from scientists during my visit to the Centre for Cullular and Molecular Biology, Hyderabad, India, in a conversation on 6 April 1989 with the director, Pushpa Bharghava, and Colin Blakemore, professor of Physiology at Oxford, who was visiting at the time. Leroy Hood, US biologist, also suggested mass infant DNA screening in the context of the human genome project, in the BBC2 programme 'The Book of Man', and it has also been mentioned in the context of DNA fingerprinting.

7

1 Peter Newmark, 'Danger of Delay for Genetic Tests', Nature, 5 June 1986, p. 557.

2 Nancy Wexler, '"Will the Circle Be Unbroken?": Sterilising the Genetically Impaired', in Aubrey Milunsky and George J Annas II (eds.) *Genetics and the Law II*, Plenum, New York, p. 319.

3 Does that sound outlandish? A statement which illustrates that genes are a major focus in scientific medicine comes from the UK's Medical Research Council: 'Molecular biology is concerned with understanding how a cell develops and works at a molecular level; this detailed knowledge is a prerequisite for understanding what happens when something goes wrong in the cell – for example, because of disease or some genetic defect. All the information by which a cell functions is encoded in its DNA . . . Genes are specialised regions of this [DNA] sequence which code for proteins that carry out all the biochemical reactions and perform many of the structural roles in a cell.' (MRC Annual Report 1984/5, p. 34).

4 Headlines from *New Scientist*, *The Times*, *Science*, and the *Independent*, 1987–1989.

5 J H Edwards, 'Judging the Unborn', inaugural lecture delivered in the University of Birmingham, 4 February 1969, published as a pamphlet; Ciril Clarke, *Human Genetics and Medicine*, Edward Arnold, London, 1977, p. 1.

6 Quoted in 'The Book of Man', BBC 2, p. 22.

7 Jill Rakusen, 'In Pursuit of the Perfect Baby', printed 1982, Women's Health and Reproductive Rights Information Centre. See list of resources.

8 Theresia Degener, unpublished paper. See also Anne Finger, *Past Due: A Story of Disability, Pregnancy and Birth*, The Women's Press, London, 1991.

9 'The Mind Machine', BBC, 18 September 1988.

10 John J Wasmuth, et al., 'A Highly Polymorphic Locus Very Tightly Linked to the Huntington's Disease Genes', *Nature*, 21 April 1988, p. 734.

11 'Chemical Clue to Huntington's Disease', *New Scientist*, 22 May 1986, p. 28.

12 Phrase used in Wasmuth, et al., 'A Highly Polymorphic Locus . . .'

13 See Steve Connor, 'Researcher Withholds Gene Probe', *New Scientist*, 13 March 1986, p. 17; John Maddox, 'Proprietary Rights to Research', *Nature*, 6 March 1986, p. 11.

14 Germaine Greer, 'Home Thoughts', *Independent Magazine*, July 1989, p. 14.

15 'Gene for Cystic Fibrosis', *New Scientist*, 14 May 1987, p. 35; 'Reverse Genetics and Cystic Fibrosis', the *Lancet*, 23 September 1989, p. 755.

16 Leslie Roberts, 'Race for Cystic Fibrosis Gene Nears End',

Science, 15 April 1988, pp. 282–5.

17 Robin McKie, 'Living with the Secrets of our Genes', the *Observer,* 10 September 1989.

18 Amino acid toxicity reported in Rif S El-Mallakh, and Daniel P Potenza, 'Bittersweet', *Science for the People,* November/December, 1988, pp. 17–18.

19 Thomson Prentice, 'Gene Link to Schizophrenia', *Times* (London), 25 July 1988; 'Faulty Hereditary Gene Found in Schizophrenia Sufferers', *Guardian,* 25 July 1988; Marjorie Wallace, 'Unlocking A Family Secret', *Times* (London), 28 July 1988; Deborah M Barnes, 'Schizophrenia Genetics A Mixed Bag', *Science* 18 November 1988, p. 1009; Michael Gill, 'Molecular Genetics and Schizophrenia', *British Medical Journal,* 3 December 1988, p. 1426.

20 *New Internationalist,* issue on Madness, July 1990, p. 22.

21 *Ibid.;* see Dorothy Rowe, *Beyond Fear,* Fontana, London, 1987.

22 *Living with Schizophrenia,* Channel 4, 29 May 1990.

23 Suman Fernando, 'The Same Difference', *New Internationalist,* issue on Madness, July 1990, pp. 24–5: 24.

24 Richard J Marshall, 'The Genetics of Schizophrenia Revisited', *Bulletin of the British Psychological Society* 37, 1984, pp. 177–81. Following quotes pp. 178, 181, 178, 179, respectively.

25 Cited in *Independent on Sunday,* 'Better Care Reduced Schizophrenia', 4 March 1990; see Michael Gill, 'Molecular Genetics and Schizophrenia', *British Medical Journal,* 3 December 1988, p. 1426.

26 Morton Hunt, 'The Total Gene Screen', *New York Times Magazine,* 19 January 1986, pp. 33–66. Following quotes pp. 33, 38.

27 Cited in *ibid.;* Mary Sue Henifin and Ruth Hubbard, 'Genetic Screening in the Workplace', *geneWATCH,* November/December 1983, pp. 5–9.

28 Quoted in Jeremy Green, 'Detecting the Hypersusceptible Worker: Genetics and Politics in Industrial Medicine', *International Journal of Health Services* 13(2), 1983, pp. 247–64; see Catherine J Damme, 'Controlling Environmental Mutagens: Through Market Incentives or State Action?', in *Genetics and the Law II,* Milunsky and Annas (eds.), p. 390.

29 Discussed in Jonathan Harwood (book review, *Race Hygiene and National Efficiency: the Eugenics of Wilhelm Schallmayer,* Sheila Faith Weiss, University of California Press, 1988), *Times Higher Education Supplement,* 1 July 1988; see Damme, *Genetics and the Law II.*

30 Quoted in Isabelle Gidley and Richard Shears, *Alzheimer's Disease: What It Is, How to Cope,* Unwin, London, 1988, p. 96.

31 'How Diagnostic DNA Meets its Match', *New Scientist,* 6 May 1989, p. 48.

32 William Bains, 'Disease, DNA and Diagnosis', *New Scientist,* 6 May 1989, pp. 48–51.

33 Quoted in Hunt, 'The Total Gene Screen', pp. 55–6.

8

1 Sue Meek, 'Reproductive Technology: Present Practices and Future Implications', in *Future Challenges for Australia: The Biotechnology Revolution*, selected papers, August 1986.

2 Michael Ashburner, 'A Shift from Soft to Hard Genetics' (book review), *Times Higher Education Supplement*, 18 May 1990.

3 'The Book of Man', BBC, p. 6.

4 Pushpa Bhargava, 'The Dawn of the Age of Biology', reprint of article appearing in *Man & Development*, December 1985, pp. 11–28. pp. 23, 25 (emphasis in original).

5 Personal communication from Nancy Marisa Gomez, Chapman College, Orange, California; Cindi McMenamin, 'Egg-donor Program Offers Hope of Birth', SBV/RSM *News*, 11 April 1990, p. A4; conversation with Dr Charanjit Babra, April 1989, National Insitute of Immunology, Delhi.

6 Joseph Fletcher, *The Ethics of Genetic Control: Ending Reproductive Roulette*, Anchor Books, Anchor Press/Doubleday, New York, 1974, p. 180.

7 For critical opinion from WHO see Marsden G. Wagner and Patricia St Clair, 'Are *In Vitro* Fertilisation and Embryo Transfer of Benefit to All?' the *Lancet*, 28 October 1989, pp. 1027–9. Most recently, Patricia A St Clair Stephenson, 'The Risks Associated with Ovulation Induction', *Iatrogenics* 1, 1991, pp. 7–16; WHO Summary Report, Consultation on the Place of *In Vitro* Fertilisation in Infertility Care, Copenhagen, 1990, issued by WHO Regional Office for Europe.

8 *Ibid.*

9 See Naomi Pfeffer and Allison Quick, *Infertility Services: a Desperate Case*, a report for the Greater London Association of Community Health Councils, 100 Park Village East, London NW1 3SR, 1988.

10 See Renate Klein and Robyn Rowland, 'Women as Test-Sites for Fertility Drugs: Clomiphene Citrate and Hormonal Cocktails', *Reproductive and Genetic Engineering: Journal of International Feminist Analysis* 1(3), 1988, pp. 251–73; Renate Klein, *Infertility: Women Speak Out About Their Experiences of Reproductive Medicine*, Pandora, London, 1989; Françoise Laborie, 'News from France and Elsewhere', *Reproductive and Genetic Engineering: Journal of International Feminist Analysis* 1(1), 1988, pp. 77–86. Marian E Carter and David N Joyce, 'Ovarian Carcinoma in a Patient Hyperstimulated by Gonadotropin Therapy for *In Vitro* Fertilisation: A case report', *Journal of In Vitro Fertilisation and Embryo Transfer* 4(2), pp. 126–8.

11 Anne McLaren, 'Reproductive Options, Present and Future', in C R Austin and R V Short (eds.), *Manipulating Reproduction*,

Cambridge University Press, Cambridge, 1988, p. 184.

12 Fletcher, *The Ethics of Genetic Control*, p. 180.

13 International Survey of Medical Genetics, presented at the Seventh Congress of Human Genetics, September 1986, Berlin, unpublished data in preparation for publication.

14 Wexler, '"Will the Circle be Unbroken?"' p. 319 Barbara von Ow, 'Eichmann Slur on Geneticists', *Times Higher Education Supplement*, 7 June 1985, referred to the controversy in Germany over the brand of humanism which embraces an ethic of genetic manipulation of people.

15 Explained by and quoted in Fletcher, *The Ethics of Genetic Control* pp. 3, 178.

16 *Ibid.*, p. 178. It should be noted that Lederberg estimated that environmental factors such as drugs, food additives and air pollution might account for 80 per cent of human gene mutations (as reported in Daniel J Kevles, *In the Name of Eugenics: Genetics and the Uses of Human Heredity*, Knopf, New York, 1985).

17 Fletcher, *The Ethics of Genetic Control*, p. 187; Wexler, '"Will the Circle Be Unbroken?"', p. 317; Thomas B Mackenzie and Theodore C Nagel, 'When a Pregnant Woman Endangers Her Foetus', *Hastings Center Report*, February 1986, pp. 24–5, followed by a commentary by Barbara Katz Rothman, p. 25.

18 Barbara Katz Rothman, 'The Products of Conception: the Social Context of Reproductive Choices', *Journal of Medical Ethics* 11, 1985, pp. 188–92; Ellen Hopkins, 'High Tech Pregnancies', The *Newsday Magazine*, 11 August 1985.

19 McLaren, 'Reproductive Options', p. 184.

20 reported in *Biotechnology Information News*, no. 24, July 1990, published by The British Library Biotechnology Information Service. There are other ways to sex animal embryos, for example, by looking at a substance on the surface of the embryo which is only produced by male embryos.

21 Bharati Sadasivam, 'The Silent Scream', the *Illustrated Weekly of India*, 14 September 1986, pp. 38–41: 38, 40, 39.

22 Forum Against Sex Determination and Sex Pre-Selection, Bombay, *Campaign Against Sex Determination and Sex-Preselection In India: Our Experience, Report*. See resources.

23 Madhu Kishwar, 'Sex Determination Tests in India', presented at the Third International Interdisciplinary Congress on Women, Dublin, 6–10 July 1987.

24 Quoted in Diana Smith, 'Science Gives Korea a New Sex Problem', *Observer* (London), 28 September 1986.

9

1 Nigel Williams, 'Revolution in the Family', the *Guardian*, 7 March 1989.

2 Audre Lorde, *Zami: A New Spelling of My Name*, Sheba, London, 1984, p. 14.

3 Quotes from, respectively, Miranda Robertson, 'The Proper Study of Mankind', *Nature*, 3 July 1986, p. 11; James Bruce Walsh, and Jon Marks, letter to the editor, *Nature*, 14 August 1986, p. 590; Christopher Joyce, "The Race to Map the Human Genome', *New Scientist*, 5 March 1987, pp. 35–9; Andrew Veitch, 'Mapping the Human World', the *Guardian*, 16 February 1988.

4 Quoted in 'The Book of Man', BBC2, p. 12.

5 *Ibid.*, p. 20.

6 *Ibid.*, p. 15.

7 Quoted in Jeremy Cherfas, 'A Guide to Being Human', *New Scientist*, 25 February 1988, pp. 30–1.

8 Arnold Pacey, *The Culture of Technology*, Basil Blackwell, Oxford, 1983, p. 81.

9 Quoted in 'The Book of Man', BBC2, p.18.

10 P M Bhargava, 'Social Implications of Modern Biology', 1988, pp. 24, 26, 27.

11 Richard Dawkins, *The Selfish Gene*, Paladin, London, 1978, p. x.

12 These examples were given in 'Tough Talk on Surrogate Birth', *Nature* 313, p. 95; Linda Birke, 'Reproductive Strategy that Does Not Add Up' (book review), *New Scientist*, 13 March 1986, p. 50; Pamela Wells, 'Why King Oedipus Was Wrong', the *Guardian*, 28 August 1987; *Today*, BBC Radio 4, 20 February 1986.

13 Quoted in 'The Book of Man', BBC, p. 3.

14 Annette Goerlich and Margret Krannich, 'The Gene Politics of the European Community', in *Reproductive and Genetic Engineering: Journal of International Feminist Analysis* 2(3), 1989, pp. 201–18: 213, 216.

15 'A Shred of Evidence', Central Television, June 1989.

16 Paul Hoyland, 'Rape Case Highlights Needs to Improve DNA Test', the *Guardian*, 10 October 1987.

17 Donna Cremans, 'DNA Fingerprinting: What's at Stake?', *geneWATCH*, January/February 1988, pp. 9–10.

18 Steve Wright, 'Tactical Technology', *Science for People*, issue 65, Autumn 1987, pp. 2–4; John Carvel, 'Met Chief Calls for Wider Use of DNA Testing', the *Guardian* (London), 21 September 1990.

19 Eric Lander, 'DNA Fingerprinting on Trial', *Nature*, 15 June 1989, pp. 501–5.

20 *Ibid.*, p. 501 (emphasis in original).

21 Reported in Steve Connor, 'Genetic Fingers in the Forensic Pie', *New Scientist*, 28 January 1988, pp. 31–2: 32.

22 Robin McKie, 'DNA Tests Black Immigrants', the *Observer*, 31 January 1988.

23 Barbara Beckwith, 'Science for Human Rights', *Science for the People*, January/February 1987, pp. 6–9, 32; this and following quotes pp. 6–7, 9.

24 Wright, 'Tactical Technology', p. 2; Trevor Pinch (book review), *Times Higher Education Supplement*, 12 May 1989.

10

1 John Hubner, 'Hidden Arms Race', *West*, 15 April 1984, p. 12.

2 Quoted in *Spare Rib*, 'Innu Women and Nato', October 1989, p. 12.

3 Susan Wright and Robert L Sinsheimer, 'Recombinant DNA and Biological Warfare', *Bulletin of the Atomic Scientists*, November 1983, p. 20.

4 Hubner, 'Hidden Arms Race', p. 13.

5 Pacey, *The Culture of Technology*, p. 174; Sir Solly Zuckerman, *Scientists and War: The Impact of Science on Military and Civil Affairs*, Hamish Hamilton, London, 1966.

6 Linda Bullard, 'Killing Us Softly: Toward a Feminist Analysis of Genetic Engineering', in Patricia Spallone and Deborah Lynn Steinberg (eds.) *Made to Order: The Myth of Reproductive and Genetic Progress*, Pergamon Press, Oxford, 1987, pp. 111, 114.

7 Leonhard S Wolfe, 'Chemical and Biological Warfare: Medical Effects and Consequences', *McGill Law Journal* 28(3), July 1983, pp. 732–49: 741.

8 Arthur H Westing, 'Chemical and Biological Weapons: Past and Present', *Peace and the Sciences*, Vienna, 1982(3), pp. 25–37. Reprinted twice in German in: K. Lohs (ed.), *Kalte Tod*, Pahl-Gugenstein, Cologne, 1982; K. Lohs (ed.), *Europa: Giftfass oder chemiewaffenfrei?*, Pahl-Gugenstein, Cologne, 1986.

9 Quoted in Bullard, 'Killing Us Softly', p. 113.

10 Steven Rose, 'Genetic Engineering – Military Implications', text of his presentation at Heidelberg Conference on the social implications and hazards of genetic engineering, 7 March 1986. Kindly provided by Alastair Hay.

11 Quoted in Hubner, 'Hidden Arms Race'.

12 Nachama L Wilker and Nancy Connell, 'Update on the Military and Biotechnology', *geneWATCH*, July/October 1987, pp. 1–3: 1.

13 Charles Pillar, 'DNA – Key to Biological Warfare?', *The Nation*, 10 December 1983, title page and p. 598.

14 Quoted in John Horgan, 'Snakebit', *Scientific American*, July 1988, pp. 6–7: 7.

15 MoD contracts reported in Steven Rose, 'Biotechnology at War', *New Scientist*, 19 March 1987, pp. 33–7.

16 Arthur Westing, 'Cultural Constraints on Warfare: Micro-organisms as Weapons', *Medicine and War* 4, 1988, pp. 85–95.

17 Bullard, 'Killing Us Softly', p. 115.; Barbara Hatch Rosenberg, 'International Biological Weapons Update', *geneWATCH*, July/October 1987, pp. 6–7, 15–16; Jonathan King, 'Resisting the Militarization of Biomedical Research', *geneWATCH*, July/October 1987, pp. 4–5.

18 Horgan, 'Snakebit', p. 6. First two cases cited in Christine Skwiot, 'The Army on Trial', *geneWATCH*, July–October 1987, pp. 13–14. The third described in Foundation on Economic Trends, press release, 'Large quantity of dangerous virus missing from army biowarfare laboratory at Ft. Detrick, former senior scientist charges army with cover-up and lax security regulations, court asked to enjoin all biowarfare research until adequate security procedures are adopted.' (See resources.)

19 Leonard Cole, 'Return to Biological Warfare Outdoor Testing?', *geneWATCH*, July–October 1987, pp. 8–11.

20 Cole, 'Return to Biological Warfare', p. 9; Rosenberg, 'International Biological Weapons Update', p. 7.

21 Wright and Sinsheimer, 'Recombinant DNA and Biological Warfare', pp. 20–6.

22 Jonathan King, "The Threat and Fallacy of a Biological Arms Race', symposium on 'Biological Research and Military Policy', American Association for the Advancement of Science, 26 May 1984. Kindly provided by Alastair Hay.

23 Protests were co-organised by the Bhopal Action Group, London Greenpeace and two groups primarily concerned with the environmental and social effects of extractive industries: Partizans, a long-standing campaigning group which seeks to minimise the damaging effects of RTZ, the world's most powerful single mining company; and Minewatch, formed by Partizans for the purpose of gathering and disseminating information on the nature of specific types of mining, with emphasis on their social impacts especially concerning the infringement of indigenous land rights. Contact Partizans, 218 Liverpool Rd, London N1 1LE.

24 Fowler, Lachkovics, Mooney and Shand, 'The Laws of Life', p. 27.

25 *Ibid.*, p. 25.

Epilogue

1 Chapter title, in Fowler, Lachkovics, Mooney and Shand, 'The Laws of Life'.

2 Jeremy Cherfas, *Man Made Life: A Genetic Engineering Primer*, Basil Blackwell, Oxford, 1982.

3 Robert Edwards and Patrick Steptoe, *A Matter of Life*, Hutchinson, London, 1980, p. 17.

4 George, *A Fate Worse Than Debt*.

5 Helen Bequaert Holmes, 'Hepatitis – Yet Another Risk of *In Vitro* Fertilisation?', *Reproductive and Genetic Engineering: Journal of International Feminist Analysis* 2(1), 1989, pp. 29–37.

6 Vandana Shiva, 'We Have Survived the White Man's Ways', *Spare Rib*, September 1989, pp. 6–10.

7 Penang Declaration in Biotechnology and Genetic Engineering, Penang, Malaysia, 9 April 1987. Kindly provided by Helga Satzinger.

8 Quoted in Benedikt Haerlin, 'Genetic Engineering in Europe', in Wheale and McNally, *The Bio-Revolution*, p. 253.

9 Fowler, Lachkovics, Mooney and Shand, 'The Laws of Life', p. 298.

Appendix

1 Gail Vines, 'Missing Genes May "Hold Back" Cancer', *New Scientist*, 28 July 1988, p. 37.

2 W French Anderson, 'Prospects for Human Gene Therapy', *Science* 226, 26 October 1984, pp. 401–9: 401.

3 CERES' evidence reprinted in *Review*, 'Regulating Genetic Manipulation', May 1990, pp. 13–20.

4 Gail Vines, 'New Tools to Treat Genetic Disease', *New Scientist*, 13 March 1986, pp. 40–42.

5 Peter Wheale and Ruth McNally, 'Technology Assessment of a Gene Therapy', *Product Appraisal* 3, (4) December 1988, pp. 199–204.

Glossary

1 Donna Haraway, 'The Biopolitics of Postmodern Bodies: Determinations of Self in Immune System Discourse', *difference* 1(1), 1988, pp. 3–43, see especially p. 10.

2 RCEP, *Thirteenth Report*, pp. 13, 126.

3 'Maternal Dysinheritance', *Scientific American*, October 1988,

p. 19, reported new genetic disease find from researchers at Emory University School of Medicine.

4 Victor McKusick, *Human Genetics*, Prentice Hall, Englewood Cliffs, NJ, USA, 1964, p. 141.

5 Plucknett, et al., *Gene Banks*, p. xiii.

6 Fowler, Lachkovics, Mooney and Shand, 'The Laws of Life', p. 304.

7 *Dictionary of Biology*, Sphere Reference, London, 1985, p. 155.

8 After RCEP, *Thirteenth Report*.

9 E O Wilson (ed.) and Frances M Peter (associate ed.), *Biodiversity*, National Academy Press, Washington DC, 1988, p. 1.

Resources

African Centre for Technology Studies, PO Box 45917, Nairobi, Kenya. Executive director Calestous Juma.

Biotechnology Business Research Group is an informal international network formed for the interchange of information and resources concerning the economic and social impacts of genetic and reproductive biotechnology. Membership and periodic newsletter is free to anyone interested. It is run under the auspices of an independent company, Bio-Information (International) Ltd, specialising in analytical information on the same. London office: 25 Northlands St, Camberwell, London, SE5 9PL; Tel: 071 737 5139. Directors Peter R Wheale and Ruth M McNally.

The British Library's Biotechnology Information Service, 25 Southampton Buildings, London WC2A 1AW, publishes *Biotechnology Information News*, subscription free; its 'Guide to Bioethics' lists many books which specifically address genetic engineering.

BBC (British Broadcasting Corporation), London, *Horizon* series programme on genetic engineering and food crops, 'Guess What's Coming to Dinner'; and in the *Split Screen* Series, 'Genetic Engineering – Is It Dangerous?'

Channel 4, London, aired 'Soft Cell', a feminist analysis of genetic engineering. For videotape copies write to the producers, Steel Bank Film Co-op, Brown St, Sheffield S1 2BS; Tel: 0742 721235.

The Council for Responsible Genetics (formerly The Committee for Responsible Genetics), 19 Garden St, MA02138, USA; Tel: 1617 868 0870. National non-profit educational organisation dedicated to raising social issues in genetics and biotechnology; publishes the bulletin *geneWATCH*.

CERES (Consumers for Ethics in Research) is a forum of individuals and organisations in Britain concerned with all aspects of biomedical research and new treatments, and mainly research into preconception, pregnancy and the first year after birth. Write to the Membership Secretary, Churchbury House, Windmill Road, London SW19 5NQ.

CEAT (Coordination Européene des Amis de la Terre), 29 rue Blanche, Brussels, Belgium; Tel 32 2 5377228. CEAT is the International Friends of the Earth clearing house on biotechnology, and can supply names and addresses of organisations in other countries.

FINRRAGE (Feminist International Network of Resistance to Reproductive and Genetic Engineering) formed in 1985 to provide an information and working network for women critically concerned with the diverse developments of reproductive and genetic technologies and their effects on women; it organises regional and international meetings. Contact the international co-ordinating group, PO Box 201903, 2000 Hamburg 20, Germany.

Forum Against Sex Determination and Sex Pre-Selection, Bombay, c/o Women's Centre, B–104 Sunrise Apts, Above Canara Bank, Vakola, Santacruz (E), Bombay 400 055, India.

Foundation on Economic Trends, 1130 17th Street NW, Suite 630, Washington DC 20036, USA; Tel: 202 466 2823.

Gen-ethisches Netzwerk e.V. (Gen-Ethic Network), Winterfeldt Str 3, Berlin 30, Germany; Tel: 49 30 215 3991. International information and communication clearing house founded in 1986 for facilitating a broad public dialogue about the objectives, applications, dangers and consequences of genetic engineering, and to discuss possible alternatives.

The Genetics Forum, 258 Pentonville Rd, London, N1 9JY; Tel 071 278 6578. National public pressure organisation which brings together concerned individuals and organisations.

GRAIN (Genetic Resources Action International), formerly the ICDA Seeds Campaign (International Coalition for Development Action). Apartado 23398, E 08080, Barcelona, Spain; Tel: 34 3 412 11 89. Publishes a citizens action guide on agricultural biotechnology.

GAB (Gruppo di Attenzione Sulle Biotechnologie), via Iglesias 33, Milan 20128, Italy; Tel: 39 2 27001135.

KSB (Kontaktgroep Biotechnologies), Studium Generale, Landbouwuniversiteit, Gen. Foulkesweg 1, 6703 BG Wageningen, The Netherlands; Tel: 31 83 70 82030.

Research Foundation for Science and Ecology, 105 Rajpur Road, Dehra Dun 2480001, India. Director Vandana Shiva.

Women's Global Network on Reproductive Rights, NWZ. Voorburgwal 32, 1012 RZ Amsterdam, The Netherlands; Tel: 31 20 209672. An autonomous network of groups and individuals in every continent. They publish a newsletter reporting on developments and campaigns.

WHRRIC(Women's Health and Reproductive Rights Information Centre), 52 Featherstone St, London EC1Y 8RT, England; Tel: 071 251 6580. Provides information on all aspects of women's health including medical and legal developments, and publishes a newsletter.

Selected Bibliography

Arditti, Rita, Renate Duelli Klein and Shelley Minden, (eds.) *Test-Tube Women: What Future for Motherhood?*, Pandora, London, 1984.

Arun, Banashree Mitra, Dr J P Jain, Jaya Srivastava, Miloon Kothari, Dr Mira Shiva, Prakash, Dr P S Sahani, Sunanda, T Krishna, Vandana Bedi, *Crime Goes Unpunished: A Report on the Cholera Epidemic*, New Delhi, October 1988. For copies contact Dr J P Jain, 4205, Budh Nagar, Tri Nagar, Delhi 110035, India.

Bains, William, *Genetic Engineering for Almost Everybody*, Penguin, Harmondsworth, 1987.

Brunner, Eric, *Bovine Somatotropin: A Product in Search of a Market*, London Food Commission, 88 Old Street, London, 1988.

Claußen, Bernhard, *Proceedings of the European Workshop on Law and Genetic Engineering*, Hamburg, 14–15 December 1989, published 1990. Available from BBU Verlag GmbH, Prinz-Albert-Str 43, 5300 Bonn 1, Germany.

Corea, Gena, *The Mother Machine*, The Women's Press, London, 1988.

Etzioni, Amitai, *Genetic Fix: the Next Technological Revolution*, Harper & Row, New York, 1973.

Finger, Anne, *Past Due: A Story of Disability, Pregnancy and Birth*, The Women's Press, London, 1991.

Fowler, Cary, Eva Lachkovics, Pat Mooney and Hope Shand, 'The Laws of Life: Another Development and the New Biotechnologies', *development dialogue*, 1–2, 1988. Available from the Dag Hammarskjöld Foundation, Övre Slottsgatan 2, S–752 20 Uppsala, Sweden.

geneWATCH 4 (4–5), July–October 1987, special issue on biotechnology and the military.

George, Susan, *A Fate Worse Than Debt*, Penguin, Harmondsworth, 1988.

Green, Jeremy, 'Detecting the Hypersusceptible Worker: Genetics and Politics in Industrial Medicine', *International Journal of Health Services* 13 (2), 1983, pp. 247–64.

Hall, Stephen S, *Invisible Frontier: The Race to Synthesize a Human Gene*, Sidgwick & Jackson, London, 1988.

Harding, Sandra, *The Science Question in Feminism*, Cornell University Press, Ithaca (USA) and London (UK), 1986.

Hatchwell, Paul, 'Opening Pandora's Box: The Risks of Genetically Engineered Organisms', *The Ecologist* 14(4), 1989, pp. 130–6.

James, Walene, *Immunization: The Reality Behind the Myth*, Bergin & Garvey, Ma, USA, 1988.

Joyce, Christopher, 'The Race to Map the Human Genome', *New Scientist*, 5 March 1987, pp. 35–9.

Juma, Calestous, *The Gene Hunters: Biotechnology and the Scramble for Seeds*, Zed Books, London, 1989.

Kevles, Daniel J, *In the Name of Eugenics: Genetics and the Uses of Human Heredity*, Knopf, New York, 1985; Penguin, Harmondsworth, 1986.

Lander, Eric, 'DNA Fingerprinting on Trial', *Nature*, (London), 15 June 1989, pp. 501–5.

McNeil, Maureen, et al. (eds.) *The New Reproductive Technologies*, Macmillan, London, 1990.

Marshall, J Richard, 'The Genetics of Schizophrenia Revisited', *Bulletin of the British Psychological Society* 37, 1984, pp. 177–81.

Marx, Jean L (ed.) *A Revolution in Biotechnology*, Cambridge University Press, Cambridge, 1989.

Mooney, Pat Roy, 'The Law of the Seed', *development dialogue*, 1–2, 1983. Available from the Dag Hammarskjöld Foundation (see under Fowler).

Murphy, Séan, Alastair Hay and Steven Rose, *No Fire, No Thunder: The Threat of Chemical and Biological Weapons*, Pluto Press, London, 1984.

Pacey, Arnold, *The Culture of Technology*, Basil Blackwell, Oxford, 1983.

Patenting Life Forms in Europe: An International Conference at the European Parliament, 7–8 February 1989, Proceedings, ICDA Seeds Campaign, (Now GRAIN), Barcelona, March 1989.

Pfeffer, Naomi and Allison Quick, *Infertility Services: A Desperate Case*, a report for the Greater London Association of Community Health Councils, 100 Park Village East, London NW1 3SR, 1988.

Pillar, Charles and Keith R Yamamoto, *Gene Wars: Military Control Over the New Genetic Technologies*, Beech Tree Books, New York, 1988, p. 195.

Plucknett, Donald, et al., *Gene Banks and the World's Food*, Princeton University Press, USA, 1987.

Rifkin, Jeremy, *Algeny*, Penguin, Harmondsworth, 1984.

Rose, Steven, R C Lewontin and Leon J Kamin, *Not In Our Genes: Biology, Ideology and Human Nature*, Penguin, Harmondsworth, 1984.

Rothman, Barbara Katz, *The Tentative Pregnancy: Prenatal Diagnosis and the Future of Motherhood*, Penguin, Harmondsworth, 1986.

Royal Commission on Environmental Pollution, *Thirteenth Report: The Release of Genetically Engineered Organisms to the Environment*, Cm 720, HMSO, July 1989.

Scutt, Jocelynne A, *The Baby Machine: Reproductive Technology and the Commercialisation of Motherhood*, Green Print, London, 1990.

Shiva, Vandana, *Staying Alive: Women, Ecology and Development*, Zed Books, London, 1988.

Sontag, Susan, *AIDS and its Metaphors*, Penguin, Harmondsworth, 1988.

Spallone, Patricia, *Beyond Conception: The New Politics of Repro-duction*, Macmillan, London, 1989. (The bibliography of this book lists many other English language feminist publications on reproductive technology from the 1980s.) And with Deborah Lynn Steinberg, (eds.) *Made to Order: The Myth of Reproductive and Genetic Progress*, Pergamon Press, Oxford, 1987. (With bibliography.)

Stanworth, Michelle, (ed.) *Reproductive Technologies: Gender, Motherhood and Medicine*, Polity Press, Cambridge, 1987.

Wagner, Marsden G and Patricia A St Clair, 'Are *In Vitro* Fertilisation and Embryo Transfer of Benefit to All?', the *Lancet*, 28 October 1989, pp. 1027–9.

Wheale, Peter and Ruth McNally, (eds.) *The Bio-Revolution – Cornucopia or Pandora's Box'*, Pluto Press, London, 1990.

   Genetic Engineering: Catastrophe or Utopia?, Harvester Wheatsheaf, Brighton; St Martin's Press, New York, 1988.

Yanchinski, Stephanie, *Setting Genes to Work: The Industrial Era of Biotechnology*, Penguin, Harmondsworth, 1985.

Glossary

> One is not born an organism. Organisms are made; they are constructs of a world-changing kind.'
> Donna Haraway[1]

amino acid – the basic units or building blocks of proteins; there are about twenty commonly occurring amino acids which join together in varying proportions to make different proteins. Commonly known examples are insulin (a protein hormone) and albumins (a group of proteins, one of which is found in egg whites).

bacteria – one-celled microscopic organisms (micro-organisms).

base – in biochemistry, the constituents of nucleic acids such as DNA; in DNA there are four of these bases called adenine, guanine, cytosine and thymidine. A gene sequence encodes genetic information, namely the ordering of amino acids in a particular protein. There are also sequences in the DNA which indicate the ordering of genes, the beginning or end of a gene, and other information.

biological diversity – the variety within and among living things and the environment which they inhabit.

biopesticide – pesticide derived from biological substances or organisms, including those derived from genetic engineering methods.

biotechnology – has many different definitions. Some define it as industrial microbiology, aimed at conversion of raw materials to a desired product, using agents of biological origin such as cells, tissue cultures, isolated enzymes. It is sometimes synonymous with genetic engineering. In its widest sense it includes all of these plus aspects of reproductive engineering, human applications, waste and pollution management, oil recovery, mineral leaching, plant breeding, diagnostics and analytical equipment, biosensors, bioelectronics, biomass energy systems, etc.

cell – 'Living organisms, from the most complex animal to the simplest microbe, are composed of cells.'[2] A cell is the basic unit of structure and function of all living things excluding viruses. Yeasts and many bacteria exist as single cells or in colonies of single cells. Plants and animals are multicellular organisms whose cells are organised into tissue and organs. A human being is composed of millions of cells. The word 'cell' comes from the Latin *cella*, 'a little room'; Robert Hooke (1625–1702) first used the word in 1655 to describe the closed cavities in oak bark which he observed under a microscope.

chimera, chimaera – an animal or plant composed of tissues that are genetically different. They can develop in nature when a mutation occurs in a cell of a developing embryo (e.g. brown patches in otherwise blue eyes). In experimental science it is a composite animal derived from fusing two or more fertilised eggs (embryos) not too distantly related, or from the union of more than one egg with one sperm. The term is also used for a DNA molecule with sequences from more than one organism.

chromosomes – the structures in cells where DNA is packaged. (DNA may be found elsewhere in a cell, but all or most of the genes are thought to be located in the chromosomal DNA. However, one research team claims to have found a new kind of genetic disease, a form of blindness which they traced to a gene on the mitochrondria, another cell component.)[3]

clones – a group of entities which are exact copies of one individual. They may be genetically identical organisms, cells or genes derivd from a common ancestor. Scientists employ a variety of techniques to create such clones.

congenital or at-birth condition – a medical condition present at birth. When a baby is born with a disability or illness, it is not necessarily a genetic inherited one. For example, Down's Syndrome is classified as a chromosomal abnormality but is not usually inherited; it is called a 'random mutation'. Other congenital conditions may be due to injury or illness during pregnancy (e.g. rubella); environmental factors such as drugs (e.g. thalidomide); exposure to chemicals (e.g. dioxin); or combinations of hereditary and environmental factors (e.g. spina bifida or some forms of cleft palate).

culture, as in tissue culture or cell culture – growing cells or portions of plants or animals under laboratory conditions, in a prepared medium (the nutrients and liquid necessary for growth).

DNA (deoxyribonucleic acid) – the chemical substance present in the nucleus of most living cells, as well as being present in bacterium cells and some viruses; the entirety of DNA is the genetic make-up of an organism. DNA is often pictured as a double helix (a screw-shaped coil) to illustrate its chemical or molecular structure. DNA is composed of two chains with nucleotides (sugar-base units) resembling links on the chain. The two chains are wound round each other to form a double helix. The double-stranded DNA can unravel to form two single chains.

embryo – in medicine, the early stages of development before recognisable human features are formed, at about eight weeks or so. Foetus is the term used from about eight weeks onward. Sometimes the word embryo refers to the earliest stages of development, especially for the time embryos can be kept alive under laboratory conditions; while the word foetus is often used for any stage after pregnancy is well established.

embryo division – splitting an embryo into two or more embryos at an early stage when each section may continue development; since both embryos have the same genetic composition, embryo division is a form of cloning.

enzyme – a protein that acts as a catalyst in biochemical reactions; an enzyme facilitates a reaction but is unchanged by it.

eugenics – the science of the improvement of the quality of the human species championed by Sir Francis Galton (1822–1911) who coined the word and proposed selective parenthood to meet the objectives of eugenics. Since then, the word and concept has been used in many different instances, and has buttressed many different social movements. In *Human Genetics*, published in 1964, geneticist Victor McKusick said that eugenics concerns itself with improvement in the genetic material. [4]

gene – specialised segment of DNA which encodes the structure of a protein. A central dogma of molecular biology states that the genetic material of all living things, save a few viruses, is in the form of DNA whose major function is to provide the basic information for protein syntesis. From that chemistry, there has been made a tremendous leap in definition, as in 'Genes, contained in living organisms, are the information blueprint for all biological life and are responsible for the characteristics of plants, animals and microbes.' [5]

gene analysis – DNA is removed from cells, spliced into pieces and the pieces are then separated from one another and analysed. Sometimes genes themselves are the subject of analysis, and sometimes other segments of DNA associated with genes.

gene library – a complete collection of the individual genes from an organism, from which a desired gene can be withdrawn and studied. A human gene library would contain all (or nearly all) types of human genes. A gene in the collection is actually cloned, that is many copies of the gene exist.

gene manipulation – used loosely by some, but specifically by geneticists to mean the formation of new combinations of DNA sequences in an organism where they do not naturally occur and in which they are capable of functioning; it entails producing DNA molecules outside the cell, and incorporating them in a virus or other system to ferry them into the organism. The ability to cross natural species barriers and place genes from one organism into an unrelated organism is one important feature of gene manipulation. See *recombinant DNA*.

gene mapping – finding the locations of genes of known functions on chromosomes, or localising any segments of DNA on the chromosomes.

gene pool – a broad category encompassing all the genes in an inter-breeding population.

gene probe or DNA probe – a piece of single-stranded DNA set to detect other DNA as a diagnostic tool (For example, a probe for haemophilia was isolated, patented and available by 1988.) It works by exploiting the structure and chemistry of DNA's double helix. The two strands of a DNA helix are held together by weak chemical forces (hydrogen bonds). If the two strands are separated, they will spontaneously come back together again. Under the right conditions, single DNA strands from different sources can re-form a double strand if they match (that is, if their building blocks are complementary). A synthetic DNA probe can be designed so that it will pair only with an exactly matched DNA, or at least one that is within 5–10 per cent of exact matching.

gene therapy – genetic engineering done on human cells; but the term has been used loosely to mean other medical methods which somewhere along the line of their development entailed gene manipulation technology.

gene transfer – refers to the introduction of a foreign gene into an organism.

genetic diversity – the variety of genes within a species, within a variety of plants or within a breed of animals.

genetic engineering – 'a technology used to alter the genetic material of living cells through direct interference with the genome in order to make them capable of producing substances or performing functions alien to the un-manipulated cell.'[6]

genetic erosion – the loss of genes from a gene pool; for example, it is used to refer to the dwindling numbers of varieties of food crops, a result of modern farming methods which limit the varieties propagated, or because of land clearing, etc.

genetic screening – the use of any of the various methods available to acquire genetic or partly genetic information about an individual.

genetics – the study of what genes are and how they work.

genome – all the genetic material contained in a cell.

genotype – the unique combination of genes in an organism; or the genetic constitution of a group of organisms. See *phenotype*.

germ plasm – a non-specific term for the total genetic information of an organism, or the total genetic variability within a population of organisms. Today it is often used to mean the genetic material. The term harks back to August Weismann's (1934–1914) theory of the continuity of the germ plasm, which proposed that the contents of the reproductive cells (egg and sperm) are inherited unchanged, and that acquired characteristics cannot be inherited.

herbicide – plant poison, more or less specific to particular plants.

in vitro – from the Latin, 'in glass', used in science to describe the growing and maintaining of cells, tissues or organs under laboratory conditions.

in vitro fertilisation, IVF – joining of egg and sperm outside the female body and under laboratory conditions. It has been used on various experimental and farm animals. It is the name given to the test-tube baby procedures when used in medical practice. A woman's eggs and man's sperm are placed in a laboratory dish in a culture medium which contains nutrients and substances necessary for growth and fertilisation. It necessitates other procedures as well, including removing eggs from a woman's ovaries, often by administering fertility and hormone drugs to the woman in order to stimulate the ovaries to produce many eggs at once.

micro-organisms – living creatures too small to be seen with the human eye, but which are visible under an optic or electron microscope; they include bacteria, viruses, protozoa (e.g. the parasite that causes malaria) and fungi (a form of plant life). Also called microbes or germs.

molecular biology – is concerned with understanding how a cell develops and works at a molecular level. The importance of genes to it is encapsulated in the statement, 'All information by which a cell functions is encoded in its DNA,' in the UK Medical Research Council's Annual Report 1985/85.

mutant – any organism affected by a mutation; in laboratory research science, the name is given to organisms whose genetic constitution is purposely changed for study.

mutation – any change in the chromosomes or DNA in cells, no matter why or how it happens, whether in nature or in the laboratory, whether occurring randomly, due to exposure to chemicals, radiation, etc., or due to direct gene manipulation. (It is an important concept in evolution theory, where mutations which occur in nature, at a low rate, affect the appearance or the behaviour of cells and organisms. Classic evolution theory states that the majority of mutations are harmful, but a tiny proportion may increase the 'fitness' of organisms; 'these spread through the population over successive generations by natural selection. Mutation is therefore essential for evolution, being the ultimate source of genetic variation.'[7]

nucleus – a structure in cells of animals and plants; it contains the chromosomes, where DNA is packaged.

organ – a collection of tissue organised for a specific function. For example, the heart and lungs are organs.

organism – any living thing (micro-organisms, plants, animals, people).

phenotype – the observable characteristics of an organism said to be determined by the interaction of its genes (its genotype) and the environment. Two organisms with the same genotype may show different phenotypes in different environments. From the Greek, *phainein*, to show; *typos*, image.

plasmids – circular molecules of DNA found in bacteria. They are used in genetic engineering to carry foreign DNA into another organism: a gene of interest can be inserted into plasmid; then the bacteria can be infected with the plasmids; if the bacteria accept the plasmids, a genetic transfer has taken place.

protein – important chemical substance in cells. Proteins are important structural elements in themselves (e.g. muscle tissue is protein), and they are also necessary for biochemical and metabolic functions (e.g. insulin is a protein in humans which is necessary for the metabolism of sugar).

protein engineering – uses genetic engineering methods to create proteins analogous to naturally occurring proteins, but altered in structure and (hopefully) improved in function. The first step in protein engineering is to clone and characterise the gene for the protein of interest.

protein synthesis – the process by which cells manufacture proteins from their constituent amino acids in accordance with the genetic information carried in the cells' DNA.

recombinant DNA – DNA that has been modified by joining together different pieces of DNA using techniques of gene manipulation rather than by traditional breeding methods;[8] recombinant DNA methods are those which allow the breaking and joining of DNA strands to produce new combinations of DNA.

restriction enzymes – protein chemicals produced by many sorts of bacteria. They 'cut' or 'splice' long DNA molecules into segments at specific points along the DNA. Discovered by scientists in the early 1970s, they were isolated and put to work for genetic engineering.

ribosome – a small spherical body within a cell that is the site of protein synthesis; usually there are many in a cell.

RNA (ribonucleic acid) – a chemical substance closely related to DNA and which plays a role in information retrieval between DNA and protein synthesis. Some viruses have RNA as their genetic material instead of DNA.

sequencing, DNA sequencing – determining the order of bases along the DNA.

sex determination – either identifying the sex of an already existing embryo or foetus, or engineering the desired sex before fertilisation occurs; the latter is often called sex predetermination. Sex selection and sex preselection are similar terms.

species – a category used to classify organisms. Generally, a species is defined as a group of similar individuals that can usually breed among themselves. Newer definitions highlight the impact of genetics on the sciences of biology and evolution, as in the definition of species as 'a population or series of populations within which free gene flow occurs under natural conditions' of breeding.[9]

tissue – a collection of a few types of cells which carry out a particular function. Examples are brain, muscle or bone tissue.

toxin – a poison produced by a living organism. The term toxin is especially used for those produced by bacteria, but can also mean other poisons, such as snake venom.

toxoid – a bacterial toxin which has been treated for use in a vaccine so that it will have lost its toxic properties but is able to stimulate an immune response. It often is used specifically in reference to the poison produced by a tetanus bacterium.

transgenic – a plant or animal into which was inserted DNA from a different species, that is, genetic material it could never have acquired through breeding in nature or through techniques used by breeders before gene manipulation technology was available.

vaccination – the introduction of a substance associated with a disease in order to stimulate an immune response which will induce immunity to the disease, and thus (hopefully) protection against it.

virus – a class of ultramicroscopic organisms capable of reproduction only within another specific living thing (bacteria, human cells, etc.); viruses are not cellular organisms, but they consist of a core of DNA or RNA surrounded by a protein coat. In gene manipulation, viruses are sometimes used to ferry genes into other cells.

Index